WEATHER STUDIES
INTRODUCTION TO ATMOSPHERIC SCIENCE

INVESTIGATIONS MANUAL

2010 - 2011 AND SUMMER 2011

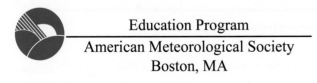

Education Program
American Meteorological Society
Boston, MA

The American Meteorological Society
Education Program

The American Meteorological Society (AMS), founded in 1919, is a **scientific and professional** society. Interdisciplinary in its scope, the Society actively promotes the development and dissemination of information on the atmospheric and related oceanic and hydrologic sciences. AMS has more than 13,000 professional members from more than 100 countries and over 175 corporate and institutional members representing 40 countries.

The Education Program is the initiative of the American Meteorological Society fostering the teaching of the atmospheric and related oceanic and hydrologic sciences at the precollege level and in community college, college and university programs. It is a unique partnership between scientists and educators at all levels with the ultimate goals of (1) attracting young people to further studies in science, mathematics and technology, and (2) promoting public scientific literacy. This is done via the development and dissemination of scientifically authentic, up-to-date, and instructionally sound learning and resource materials for teachers and students.

AMS Weather Studies, a component of the AMS education initiative since 1999, is an introductory undergraduate meteorology course offered partially via the Internet in partnership with college and university faculty. **AMS Weather Studies** provides students with a comprehensive study of the principles of meteorology while simultaneously providing classroom and laboratory applications focused on current weather situations. It provides real experiences demonstrating the value of computers and electronic access to time-sensitive data and information.

Developmental work for **AMS Weather Studies** was supported by the Division of Undergraduate Education of the National Science Foundation under Grant No. DUE - 9752416.

This project was supported, in part,
by the
National Science Foundation
Opinions expressed are those of the authors and
not necessarily those of the Foundation

Weather Studies: Investigations Manual 2010 - 2011 and Summer 2011
ISBN-10: 1-878220-08-X
ISBN-13: 978-1-878220-08-0

Copyright © 2010 by the American Meteorological Society

Cover photograph © Alamy Limited, www.alamy.com

Welcome to *AMS Weather Studies*

You are about to experience the excitement of real-world weather. This ***Weather Studies Investigations Manual*** is designed to introduce you to tools that enable you to explore, analyze, and interpret the workings of Earth's atmosphere.

This ***Investigations Manual*** is self-contained. Investigations draw from actual weather events to assist the learner in achieving their stated objectives. The investigations continually build on previous learning experiences to help the learner form a comprehensive understanding of the Earth system's atmospheric environment.

Additionally, case studies of current or recent atmospheric events are prepared in real time twice per week during fall and spring semesters in a schedule aligned with the ***Investigation Manual***'s table of contents (for Investigations 1A through 12B). These "Current Weather Studies" appear on the course website by noon, Eastern Time, on Monday and Wednesday for optional use by institutions operating on or near the AMS delivery timetable. They may be used as an alternative to the ***Manual's*** *Applications* portion of each activity or as a supplement to that printed section. Current Weather Studies accumulate each semester and remain available via a website archive. Studies expanding on ***Manual*** Investigations 13A through 15B are posted to the website at the beginning of each fall semester and available throughout the year.

Getting Started:

1. Your course instructor will provide you with the specific requirements of the course in which you are enrolled.

2. Your course instructor will provide you with the ***Weather Studies*** course website address. Record that address: ***http://***_____.

3. When the page comes up, add this address to your list of bookmarks or favorites for future retrievals. Type the login ID and password provided by your instructor when prompted for full access to the contents of the page.

   > ***Login ID:*** _____
   > ***Password:*** _____

4. Explore the course website, noting its organization and the kinds of information provided. Throughout the year, 7 days a week, 24 hours a day, the meteorological products displayed are the latest available. You will learn to interpret and apply many of these products via the ***Weather Studies*** investigations.

5. Complete ***Investigation Manual*** activities and other course requirements, including use of Current Weather Studies, as directed by your instructor.

6. **Keep Current!** Keep up with the weather and your weather studies. Weather makes more sense if you watch it in action. Visit the ***Weather Studies*** website at least once a day if you can; more often when the weather is changing.

AMS Weather Studies Investigations

1A   SURFACE AIR PRESSURE PATTERNS
* Draw isobars on a surface weather map and interpret isobar patterns.

1B   AIR PRESSURE AND WIND
* Apply the hand-twist model to surface winds in highs and lows.

2A   SURFACE WEATHER MAPS
* Decode symbols on a surface weather map and interpret weather conditions.

2B   THE ATMOSPHERE IN THE VERTICAL
* Plot a sounding on a Stüve diagram and compare to the U.S. Standard Atmosphere.

3A   WEATHER SATELLITE IMAGERY
* Compare visible and infrared satellite images for weather interpretation.

3B   SUNLIGHT THROUGHOUT THE YEAR
* Describe variations in solar radiation throughout the year by latitude.

4A   TEMPERATURE AND AIR MASS ADVECTION
* Draw isotherms on a surface map and determine areas of warm and cold air advection.

4B   HEATING AND COOLING DEGREE-DAYS AND WIND CHILL
* Calculate heating and cooling degree-days and determine wind chill.

5A   AIR PRESSURE CHANGE
* Use a meteogram to describe changes in air pressure and other weather conditions with the passage of a warm front and a cold front.

5B   ATMOSPHERIC PRESSURE IN THE VERTICAL
* Use the pressure block concept to demonstrate the influence of air density and air temperature on changes in air pressure with altitude.

6A   CLOUDS, TEMPERATURE, AND AIR PRESSURE
* Use cloud-in-a-bottle demonstration and a sounding on a Stüve diagram to illustrate how temperature changes are related to pressure changes.

6B   RISING AND SINKING AIR
* Use a Stüve diagram to illustrate dry and saturated adiabatic processes as air parcels ascend and descend in the atmosphere.

7A   PRECIPITATION PATTERNS
  * Locate and track areas of precipitation using weather radar operating in the reflectivity mode.

7B   DOPPLER RADAR
  * Describe the wind pattern detected by Doppler weather radar for a severe weather situation.

8A   SURFACE WEATHER MAPS AND FORCES
  * Examine the influence of forces on horizontal air motion near the Earth's surface.

8B   UPPER-AIR WEATHER MAPS
  * Describe the properties of a 500-mb map analysis and identify highs, lows, ridges, and troughs.

9A   WESTERLIES AND THE JET STREAM
  * Examine upper-air westerly wave patterns, the jet stream, and how these features influence midlatitude surface weather.

9B   ¡EL NIÑO!
  * Describe atmospheric and oceanic conditions that accompany periodic warmings of the tropical Pacific Ocean.

10A   THE EXTRA-TROPICAL CYCLONE
  * Describe weather conditions surrounding the center of a typical extra-tropical cyclone in the midlatitudes.

10B   EXTRA-TROPICAL CYCLONE TRACK WEATHER
  * Compare weather conditions on either side of an extra-tropical cyclone in the midlatitudes.

11A   THUNDERSTORMS
  * Examine thunderstorms as they appear on visible, infrared, and water vapor satellite images.

11B   TORNADOES
  * Determine some of the characteristics of two intense tornadoes.

12A   HURRICANES
  * Plot a hurricane as it approaches a coastal area and assess the potential threats to life and property.

12B   HURRICANE WIND SPEEDS AND PRESSURE CHANGES
  * Explore the relationships between central sea-level pressures and wind speeds throughout the life of a hurricane.

13A **WEATHER INSTRUMENTS AND OBSERVATIONS**
  * Explore the data provided by the Automated Surface Observing System (ASOS) and access weather observations for the U.S. and the world via the Internet.

13B **WEATHER FORECASTS**
  * Describe the general elements of a weather forecast and explore the NWS office forecast made available for the public.

14A **OPTICAL PHENOMENA**
  * Describe interactions of light with atmospheric water droplets and ice crystals and the resulting optical phenomena.

14B **ATMOSPHERIC REFRACTION**
  * Describe how refraction of light varies with solar altitude and how it affects periods of daylight.

15A **VISUALIZING CLIMATE**
  * Portray statistical climate values on a climograph and compare climographs from various locations to explore climate controls.

15B **LOCAL CLIMATIC DATA**
  * Interpret data appearing in the *Local Climate Data, Annual Summary With Comparative Data* and determine how to access archived data.

Investigation
1A:

SURFACE AIR PRESSURE PATTERNS

Objectives:

Much of the weather we experience arises from air put into motion because of differences in air pressure over distance. Across a horizontal surface such as at sea level, whenever and wherever the air pressure varies from one place to another the wind tends to blow from locations with relatively high air pressure towards where the air pressure is relatively low. Knowing the patterns of pressure across the country and the air motion they produce is basic to understanding the weather taking place and predicting what the weather is likely to be. By analyzing the reported values of air pressure across the country, the locations of centers of high and low pressure can be identified on weather maps. The centers typically coincide, respectively, with fair and stormy weather systems.

After completing this investigation, you should be able to:

- Draw lines of equal pressure (isobars) to show the patterns of surface air pressures across the nation at map time.
- Locate regions of relatively high and low air pressures on the same surface map.

Introduction:

Air pressure at any point on Earth's surface or in the atmosphere is equal to the weight of the atmosphere <u>above</u> that point acting on a unit area. This means that air pressure decreases with increasing altitude. Hence, the higher the elevation of Earth's surface, the lower the surface air pressure at that location. Consequently, locating centers of high and low atmospheric pressure which help to identify weather systems, requires analysis of air pressure values determined at numerous locations at the same elevation.

Air pressures routinely reported on surface weather maps are values "corrected" to sea-level. That is, air pressure readings are adjusted to what they would be if all the reporting stations were actually located at sea-level. Adjustment of air pressure readings to the same elevation removes the influence of Earth's topographical relief on air pressure readings. This adjustment allows determinations of horizontal pressure differences and recognition of pressure patterns. These patterns reveal existing broad-scale pressure areas that have a major influence on the weather.

Horizontal air pressure patterns on a weather map are revealed by drawing lines representing equal pressure. These lines are called *isobars* because every point on the same line has the same air pressure (barometric) value. Each isobar separates stations reporting higher pressures from stations with lower pressures than that of the isobar.

The **Figure 1** surface map shows air pressure in millibar (mb) units at various locations.

[One millibar (the pressure unit commonly used for atmospheric pressure) is equal to one hectopascal (hPa).] (Average midlatitude, sea-level air pressure is 1013.25 mb.) On the map, consider each pressure value to have been observed at the center of the plotted number.

1. On the Figure 1 map, the lowest plotted pressure is 1007 mb and the highest plotted value is [(*1011*)(*1016*)(*1026*)] mb.

The 1024-mb and part of the 1020-mb isobars have been drawn. **Complete the pressure analysis by finishing drawing the 1020-mb isobar. Then place the 1016-, 1012- and 1008-mb isobars.** Label each completed isobar by writing the appropriate pressure value at its ends as shown.

2. By U.S. convention, isobars on surface weather maps are usually drawn using the same interval (the difference in air pressure between adjacent isobars) as that used on the Figure 1 map. The isobar interval is [(*2*)(*3*)(*4*)(*5*)] mb. The isobar interval is selected so as to provide what is generally the most useful resolution of the field of data; too small an interval (for example, 1 mb) would clutter the map with too many lines and too great an interval (for example, 10 mb) would ordinarily mean too few lines to adequately define the pattern.

3. Also by U.S. convention, isobars drawn on surface weather maps are a series of values that, when divided by 4, produce whole numbers (e.g., $1000 \div 4 = 250$). The progression of isobaric values can be found by adding 4 sequentially to 1000 and/or subtracting 4 sequentially from 1000 until the full range of pressures reported on the map can be evaluated. Which of the following numbers would <u>not</u> fit such a sequence of isobar values: [(*1000*)(*1004*)(*1006*)(*1008*)(*1012*)(*1016*)]?

4. The change of pressure over distance is called the ***pressure gradient***. On surface weather maps, the directions of the pressure gradients (most rapid pressure change over distance) are always oriented perpendicular to the isobars. And, the closer the isobars appear on a map, the stronger the pressure gradients. From the isobar pattern you have shown to exist on the Figure 1 map, the horizontal pressure gradient is stronger across [(***West Virginia***)(***Alabama***)].

Tips on Drawing Isobars: Keep the following "rules" about drawing isobars in mind whenever you are analyzing air pressure values reported on a surface weather map.

    a. Always draw an isobar so that air pressure readings greater than the isobar's value are consistently on one side of the isobar and lower values are on the other side.

    b. When positioning isobars, assume a steady pressure change between neighboring stations. For example, a 1012-mb isobar would be drawn between 1010 and 1013 about two-thirds the way from 1010.

    c. Adjacent isobars tend to look alike. The isobar you are drawing will generally align with the curves of its neighbors because horizontal changes in air pressure from place to place are usually gradual.

d.   Continue drawing an isobar until it reaches the boundary of the plotted data or "closes" to form a loop by making its way to its starting point.

e.   Isobars never stop or end within a data field, and they never fork, touch or cross one another.

f.   Isobars cannot be skipped if their values fall within the range of air pressures reported on the map. Isobars must always appear in sequence; for example, there must always be a 1000-mb isobar between the 996-mb and 1004-mb isobars even if corresponding numbers are not plotted.

g.   Always label all isobars.

Optional: If you are unsure about your isobar-drawing skills or just crave experiences drawing isopleths (lines of constant value, including isobars) before attempting analyses of real-world weather maps, go to: *http://cimss.ssec.wisc.edu/wxwise/contour/*. Try it, you will like it.

Please note that the Internet addresses appearing in this Investigations Manual can be accessed via the "Learning Files" section of the course website. Click on "Investigations Manual Web Addresses." Then, go to the appropriate investigation and click on the address link. We recommend this approach for its convenience. It also enables AMS to update any website addresses that were changed after this Investigations Manual was prepared.

As directed by your course instructor, complete this investigation by either:

1. *Going to the Current Weather Studies link on the course website, or*
2. *Continuing to the Applications section for this investigation that immediately follows in this Investigations Manual.*

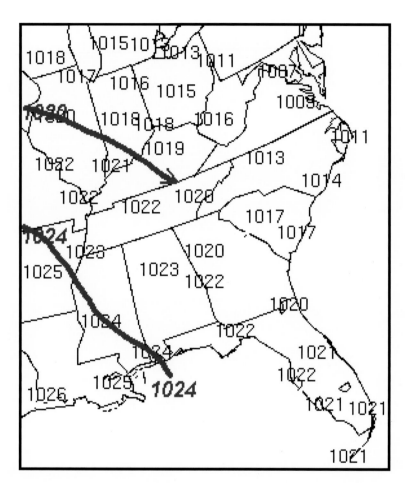

Figure 1.
Surface weather map with pressures reported in whole millibar units.

Investigation 1A: Applications

SURFACE AIR PRESSURE PATTERNS

Figure 2. Pressures was acquired from the course website and reports surface air pressures (corrected to sea level) rounded to the nearest whole millibar at 12Z 24 JAN 2010, actually 7 AM EST (6 AM CST, 5 AM MST, 4 AM PST, etc.) on Sunday morning.

5. The lowest plotted air pressure on the map is [(*976*)(*988*)(*1000*)] mb at Peoria in central Illinois.

6. The highest reported pressure is 1027 mb in [(*Del Rio, Texas*)(*Ely, Nevada*)].

7. The isobars in the conventional series that will be needed to complete the pressure analysis between the lowest and highest values on this map are:
    [(*994, 998, 1002, 1006, 1010, 1014, 1018, 1022*)
    (*999, 1003, 1007, 1011, 1015, 1019, 1023*)
    (*992, 996, 1000, 1004, 1008, 1012, 1016, 1020, 1024*)].

When an isobar value of pressure occurs at a single point and all surrounding pressures are either less than or greater than that value, the isobar need not be drawn. More than one isobar of the same value may need to be drawn if pressure values located in separate sections of the map area require it. Two isobars have already been drawn in the central U.S. where values reported on the map enclosed an extreme of the pressure pattern on the map. Also, a 1024-mb isobar has been drawn in the northeastern portion of the map area where that value was found.

Using a pencil, follow the steps below to complete the pressure analysis to determine the pressure pattern that existed at the time the observations were made. Consider each pressure value to be located at the center of the reported number.

8. Arbitrarily we might start by drawing the 1000-mb isobar continuing the series of the two central U.S. isobars already shown. This isobar may be started by entering the plotted map area in North Dakota (between 1004 and 999) and going through the "1000" located in South Dakota. Continue southward to northeastern Texas where the values then turn eastward. Continue around the already identified lower values until heading northward to exit the map area over Lake Superior. Label this isobar with its value, *1000*, at the ends of the isobar line where it enters or leaves the map field. The pressure values across the north-central U.S. within your loop of the 1000-mb isobar including the two isobars already drawn there are [(*less than*)(*equal to*)(*greater than*)] 1000 mb.

Continue drawing and labeling isobars of the series where they existed within the data pattern. Note that several of the isobar values appear in both the eastern U.S. and the western

U.S. Label your isobars with their values at the ends that extend just beyond the data field. Label the lowest value in the map area with a bold **L** and the highest value with a bold **H** (each about 1 cm high).

9. **Figure 3** is the analyzed surface pressure map from the National Oceanic and Atmospheric Administration's (NOAA) National Centers for Environmental Prediction (NCEP) for 12Z 24 JAN 2010. The Image 2 map shows the location of air pressure system centers and fronts. The Figure 3 map [(*is*)(*is not*)] the same time and date as the Figure 2 map of pressures you have just analyzed.

Compare your isobar pattern with that drawn on the Figure 3 map. The Figure 3 map is constructed by a computer based on a much more complete set of pressure values. (This may account for some of the variations between your analysis and that by the computers. It also is the source of several of the plotted Hs and Ls denoting locally minimally higher or lower pressure centers, respectively.)

10. The bold red, blue and purple lines are fronts that are generally associated with Lows (L). Light blue, green, yellow and red shadings scattered across the map are areas where radar beams were reflected back to the radar sites indicating varying intensities of precipitation (see scale to the left margin). Generally, the radar-indicated precipitation areas [(*are*) (*are not*)] associated with the Ls and frontal positions.

By analyzing the pressure values reported on weather maps to find pressure patterns, one can locate the centers of locally highest and lowest pressures, respectively. We will see that these pressure centers often mark the midpoints of major weathermakers, either regions of fair weather or stormy conditions.

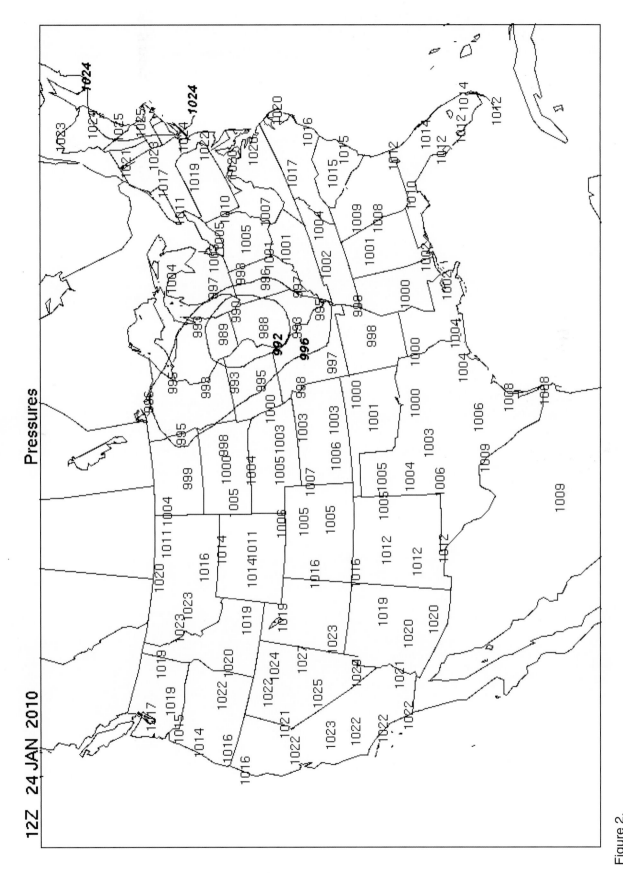

Figure 2.
Partially completed map of sea-level air pressures for 12Z 24 JAN 2010.

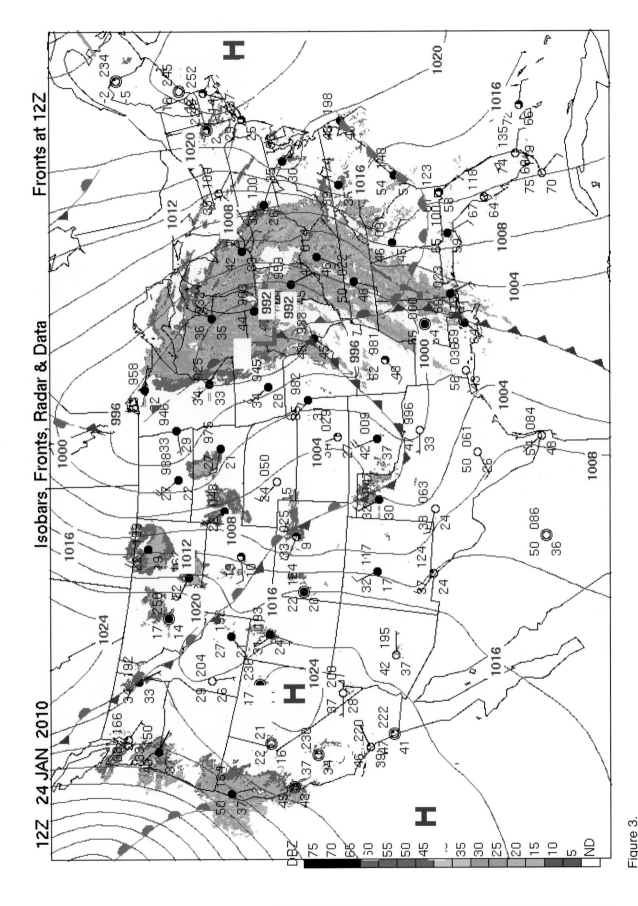

Figure 3.
Analyzed NCEP surface weather map for 12Z 24 JAN 2010 showing weather systems and isobars.

Investigation 1B:

AIR PRESSURE AND WIND

Objectives:

Air pressure, determined by the weight of the overlying air, varies from place to place and over time. Air moves in response to horizontal differences in air pressure, setting the stage for much of the weather we experience. Wind (air in motion) tends to blow from where the air pressure is relatively high to where the air pressure is relatively low. Once air is in motion, its speed and direction may be influenced by the rotation of the Earth on its axis (the Coriolis Effect) and/or contact with Earth's surface (friction). The Coriolis Effect is important in large-scale weather systems (highs and lows of weather maps, for example) and friction affects winds blowing close to the Earth's surface below an altitude of about 1000 meters.

After completing this investigation, you should be able to:

- Describe the relationship between the patterns of relatively high and low air pressure areas (Lows or **L**s and Highs or **H**s) on a surface weather map and the direction of surface winds.
- Apply the "hand-twist" model of wind direction to the circulation in actual highs and lows.

Introduction:

Turn to **Figure 1. Low**. Lightly draw a circle an inch or so in diameter around the large "L" shown on the map. The "L" marks the location of lowest pressure in a low-pressure area. Using your left hand (if you are right-handed) or your right hand (if you are left-handed), cover the circle with your palm as shown to the right.

[Note: The following analysis is more easily conducted if standing up.]
Practice rotating your hand <u>counterclockwise</u> as seen from above while gradually pulling in your thumb and fingertips as your hand turns until they touch the circle. Be sure the map does not move. Practice until you achieve a maximum twist with ease.

Place your hand back in the spread position on the map. Mark and label the positions of your thumb and fingertips 1, 2, 3, 4, and 5, respectively.

Slowly rotate your hand <u>counterclockwise</u> while gradually drawing in your thumb and fingertips. Stopping after quarter turns, mark and label (1 through 5) the positions of your thumb and fingertips. Continue the twist until your thumb and fingertips are on the circle. Connect the successive numbered positions for each finger and your thumb using a smooth curved line. Place arrowheads on the end of the lines to show the directions your fingertips and thumb moved. **The spirals represent the general flow of surface air that occurs in a typical low-pressure system.**

Now turn to **Figure 2. High**. Lightly draw a circle an inch or so in diameter around the large "H" appearing on the map. The "H" represents the location of highest pressure in a high-pressure area.

Place the map flat on your desk. With your non-writing hand, bring the thumb and fingertips of your hand close together and place them on the circle you drew as in the sketch to the right.

Rotate your hand slowly <u>clockwise</u>, as seen from above, and gradually spread out your thumb and fingertips as your hand turns. Be sure the map does not move. Practice this motion until you achieve as full a twist as you can comfortably. Place your thumb and fingertips back in the starting position on the circle. Mark and label the positions of your thumb and fingertips 1, 2, 3, 4, and 5, respectively.

Slowly rotate your hand <u>clockwise</u> while gradually spreading your thumb and fingertips. Go through about a quarter of your twisting motion. Stop, mark, and label (1 through 5) the positions of your thumb and fingertips on the map. Follow the same procedure in quarter steps until you complete a full twist.

Connect the successive numbered positions for each finger and your thumb using a smooth curved line. Place arrowheads on the ends of the lines to show the directions your thumb and fingertips moved. **The spirals represent the general flow of surface winds that occurs in a typical high-pressure system**.

1. Which of the following best describes the surface wind circulation around the center of a low-pressure system (as seen from above)?
    [(*__counterclockwise and outward spiral__*)(*__counterclockwise and inward spiral__*)
    (*__clockwise and outward spiral__*)(*__clockwise and inward spiral__*)].

2. Which of the following best describes the surface wind circulation around the center of a high-pressure system (as seen from above)?
    [(*__counterclockwise and outward spiral__*)(*__counterclockwise and inward spiral__*)
    (*__clockwise and outward spiral__*)(*__clockwise and inward spiral__*)].

3. On your desk, repeat the hand twists for the low- and high-pressure system models. Note the vertical motions of the palm of your hand. For the Low, the palm of your hand [(*__rises__*)(*__falls__*)] during the rotating motion.

4. In the case of the High, the palm of your hand [(*__rises__*)(*__falls__*)] during the rotating motion.

5. Imagine that the motions of your palms during these rotations represent the directions of vertical air motions in Highs and Lows. Vertical air motion in a Low is therefore [(*__upward__*)(*__downward__*)].

6. In the case of the High, vertical air motion is [(*__upward__*)(*__downward__*)].

7. Considering the complete air motions of the low-pressure system, air flows
[(*downward and outward in a clockwise spiral*)
(*downward and inward in a counterclockwise spiral*)
(*upward and outward in a clockwise spiral*)
(*upward and inward in a counterclockwise spiral*)].

8. In a high-pressure system, air flows
[(*downward and outward in a clockwise spiral*)
(*downward and inward in a counterclockwise spiral*)
(*upward and outward in a clockwise spiral*)
(*upward and inward in a counterclockwise spiral*)].

As directed by your course instructor, complete this investigation by either:

1. *Going to the Current Weather Studies link on the course website, or*
2. *Continuing to the Applications section for this investigation that immediately follows in this Investigations Manual.*

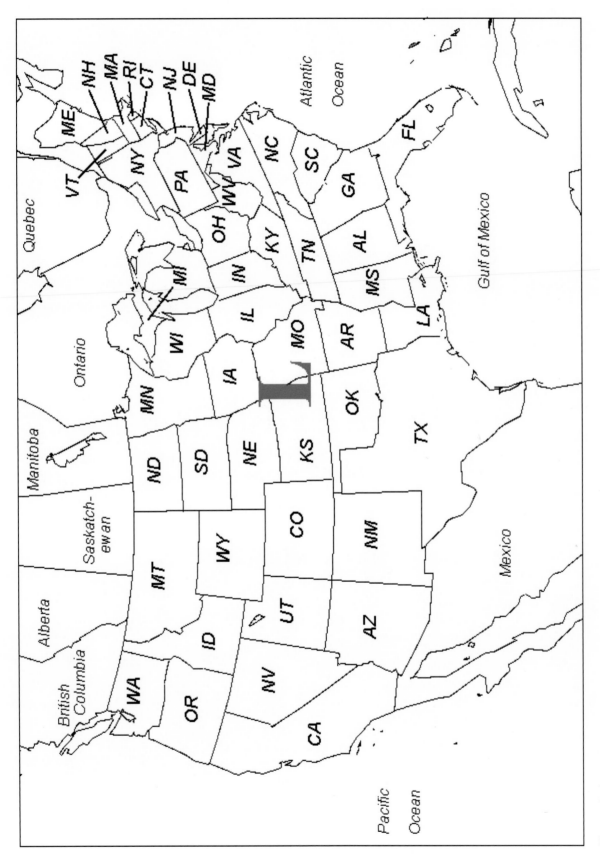

Figure 1. Low

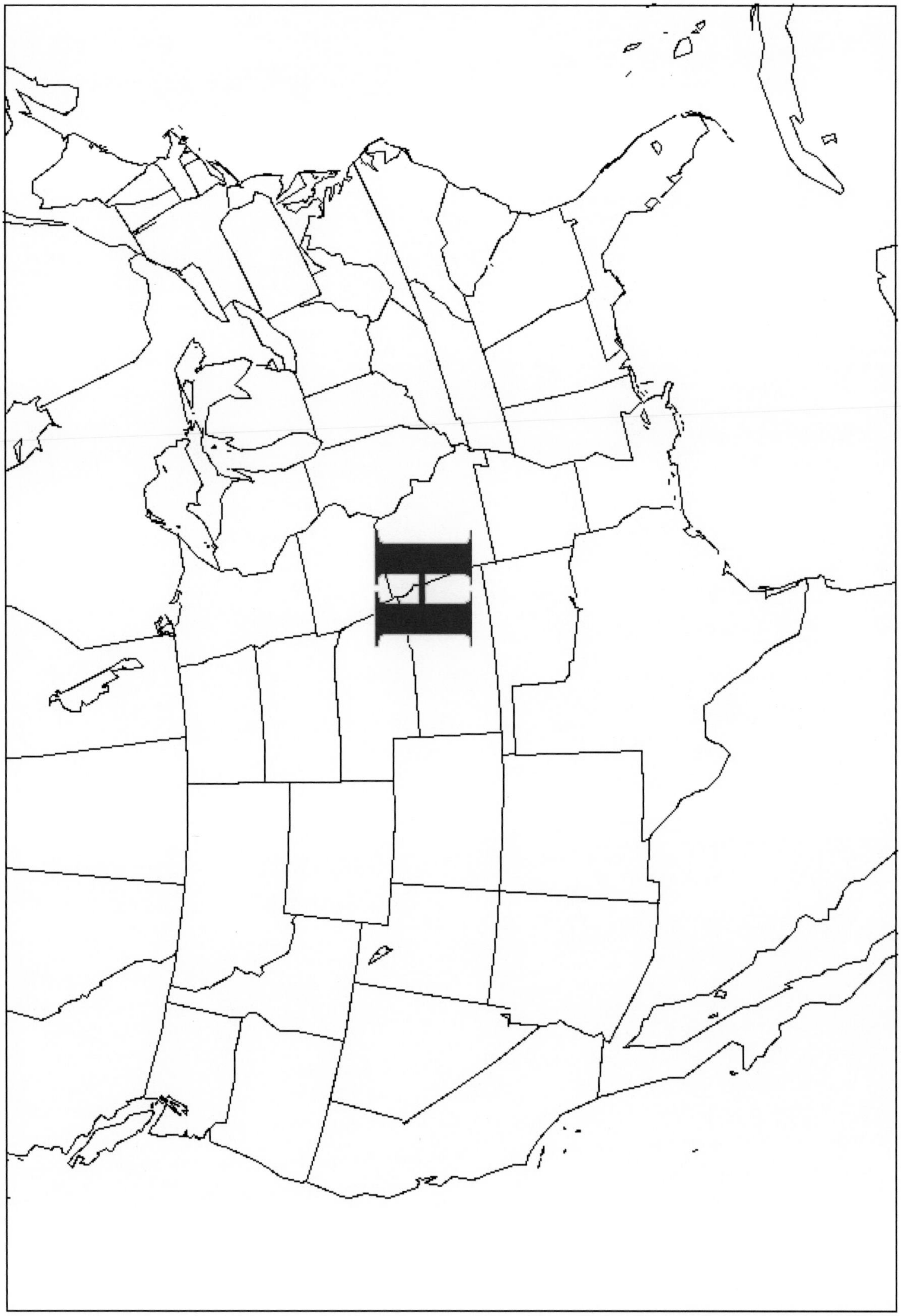

Figure 2. High

Investigation 1B: Applications

AIR PRESSURE AND WIND

Figure 3. "**U.S. - Data**" acquired from the course website is the depiction of weather conditions at stations across the contiguous U.S. at 12Z 24 JAN 2010. [12Z is five hours ahead of Eastern *Standard* Time (EST), so the map depicts conditions at 7 AM EST (6 AM CST, 5 AM MST and 4 AM PST) on Sunday morning.] The general weather conditions across the U.S. at map time were storminess stretching from the Great Lakes to the Gulf of Mexico while much of the western half of the country was cool with more showers along the West Coast.

9. Weather data at individual locations are plotted in a coded format called the "station model." The wind directions at reporting stations on the map are shown by the line (which can be thought of as an arrow shaft) which depicts the air flow <u>into</u> circles representing station locations. Wind at a station is named by the direction <u>from</u> which the air flows, *i.e.*, air arriving at the station from the north is a **north** wind. The wind direction at Bismarck, near the center of North Dakota, at map time was <u>from</u> the [(***northeast***)(***northwest***)(***southeast***)(***southwest***)].

    (All reporting surface weather stations can be identified from the "Available Surface Stations" link on the course website and identities given in the "User's Guide." Also a map of National Weather Service offices can be found at: *http://www.wrh.noaa.gov/wrh /forecastoffice_tab.php*)

10. Given the direction the wind at Bismarck was from, it would be reported as a [(***northeast***)(***northwest***)(***southeast***)(***southwest***)] wind.

    The wind speed is given by a combination of long (10 knots) and short (5 knots) "feathers" on the direction shaft. [The station model will be explained in Investigation 2A. Further details for deciphering station data can be found in your User's Guide (linked from the course website).] At map time, Bismarck had a 20-knot wind (two long feathers). [A double circle without a direction shaft signifies calm conditions, such as San Francisco, Fresno and San Diego, CA, and a shaft without feathers denotes 1-2 knots. One knot (nautical mile per hour) is about 1.2 land (statute) miles per hour.]

11. A bold red "**L**" and a bold blue "**H**" are marked on the map. Compare the *hand-twist* model of a Low to the wind directions in the several state area about this low-pressure center. Wind directions at stations across this area from Missouri to Michigan and Minnesota to Tennessee show that, as seen from above, the air spiraled generally [(***clockwise***)(***counterclockwise***)] around this low-pressure center, denoted by the **L**.

12. The air also spiraled generally [(***inward toward***)(***outward from***)] the low-pressure center.

13. This wind flow pattern about the Low is [(*consistent with*)(*contrary to*)] the *hand-twist* model of a Low.

14. The local coverage of the sky by clouds at a station is denoted by the shading within the circle representing the station. A dark circle means overcast conditions, i.e. completely cloudy sky. An open circle means clear skies. Partial shading represents the fraction of sky covered by clouds. The skies in the several state area about the Low were generally [(*clear*)(*cloudy*)].

15. The hand-twist model of a Low includes vertical motions with air rising. Based on the Low shown on this map, areas of rising air are likely to be locations of [(*clear*)(*cloudy*)] skies.

16. This pattern of cloud cover [(*was*)(*was not*)] consistent with low-pressure systems being characterized as "stormy", implying extensive cloudiness and possibly precipitation. The weather at map time at St. Louis, MO, Chicago, IL, and Detroit, MI was reported as light rain.

17. Wind directions at stations from Idaho to southern California show that, as seen from above, the air spiraled generally [(*inward toward*)(*outward from*)] the high-pressure center, denoted by the **H** in central Nevada. This flow is consistent with the *hand-twist* model of a High. However, with this still-developing High and the constraints imposed by the mountainous region, the air does not evidence the expected clockwise circulation around the high-pressure center.

18. The hand-twist model of a High includes vertical motions with air sinking. Based on stations surrounding the High on this map, especially noticeable southwest of the high-pressure center, areas of sinking air are likely to be locations of generally [(*clear*)(*cloudy*)] skies. Some stations may have cloudy skies based on air forced to rise by locally mountainous terrain that is not due to the overall weather pattern.

    The pattern of clear and partly cloudy skies in the western U.S. associated with the **H** was generally consistent with high-pressure systems being "fair" weather implying few or no areas of precipitation.

19. Look at the Figure 3 map from Investigation 1A, **Applications**. This surface weather map [(*was*)(*was not*)] for the same day and time as the U.S. - Data map of this activity.

20. Note the precipitation areas as denoted by the radar echoes on the Figure 3 map from Investigation 1A, **Applications**. The precipitation areas across the eastern half of the U.S. [(*do*)(*do not*)] further support the indication of the Low as a "stormy" weather system.

When the current weather map available on the AMS Weather Studies website shows centers of stormy Lows or fair weather Highs near your location, you might try to fit your local wind

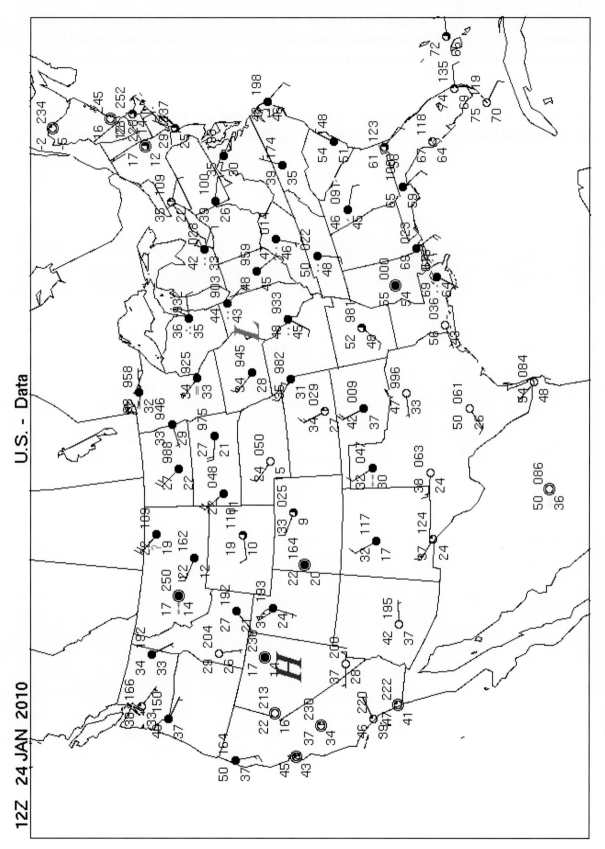

Figure 3. "U.S. - Data" map for 12Z 24 JAN 2010.

direction (as shown by a flag, for example) with map circulations and the hand-twist model of weather systems. The designation of the Ls and Hs as centers of stormy and fair weather systems, respectively, can be compared to satellite views showing clouds across the U.S. Check to see if the region immediately around an L is generally cloudy or the broad area centered on an H as mostly clear.

One tool for wind speed conversions between miles per hour and knots (as well as other quantities) and their formulae can be found at: *http://www.srh.noaa.gov/epz/?n=wxcalc*.

Investigation 2A:

Objectives:

Weather is the state of the atmosphere at a particular time and place, mainly with respect to its impact upon life and human activity. Weather is defined by various elements including air temperature, humidity, cloudiness, precipitation, air pressure, and wind speed and direction. The surface weather map is a useful tool for depicting weather conditions over broad areas.

After completing this investigation, you should be able to:

- Decode the symbols appearing on a surface weather map and describe weather conditions at various locations.
- Identify fronts appearing on the map, the weather likely to be occurring on either side of a front, and the motion of fronts.
- Describe general relationships between wind patterns and the high and low air pressure centers shown on weather maps.

Introduction:

1. Examine the surface weather map presented in **Figure 1** of this investigation. The weather map symbols shown are those commonly seen on television and in newspapers. The H's and L's identify centers of relatively high or low air pressure compared to their surroundings. Moving outward horizontally in any direction from the blue H positioned in Texas, air pressure would [(*__increase__*)(*__decrease__*)].

2. Moving outward horizontally in any direction from the red L located in Lower Michigan, air pressure would [(*__increase__*)(*__decrease__*)].

3. The thick curved lines with triangles (spikes) and/or semi-circles on the map are air mass boundaries. In the atmosphere, broad expanses of air with generally uniform temperature, humidity, and density come in contact with other masses of air having different temperature, humidity, and density characteristics. Because air masses of different densities do not readily mix, boundaries separating air masses tend to remain distinct and persistent. These boundaries, called *fronts*, typically separate warm and cold air. The leading edge of advancing cold air is a *cold front* and, as shown in the map legend in Figure 1, is signified by blue spike symbols which are pointing in the direction toward which the cold front is moving. The leading edge of advancing warm air is a *warm front* and is signified by red semi-circles on the side of the front's movement. The front plotted in the Southeastern U.S. is a [(*__cold__*)(*__warm__*)] front.

4. According to the map, persons living in South Carolina can expect [(*__colder__*)(*__warmer__*)] weather after the front passes.

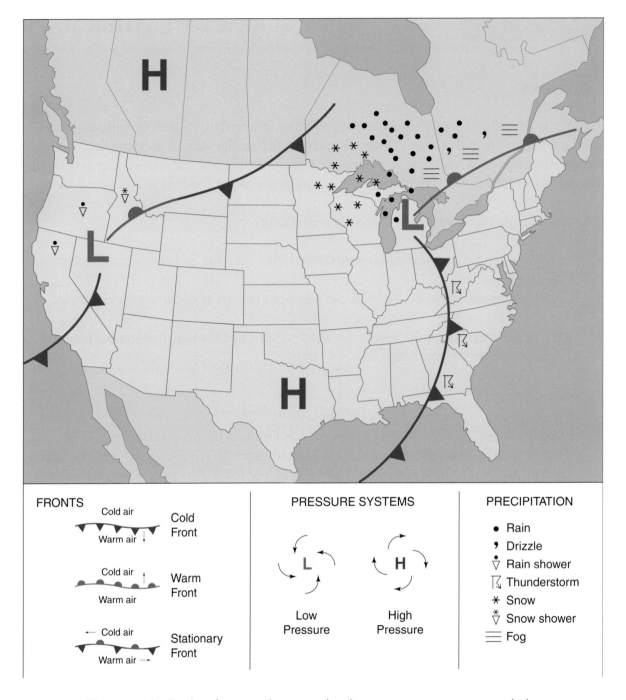

Figure 1. Idealized surface weather map showing some common map symbols.

5.   Precipitation is often depicted on weather maps by a variety of symbols as shown on this map including stars or asterisks (*) to represent **[(*rain*)(*snow*)]**.

6.   Two or three, whole or broken, horizontal lines symbolize **[(*hail*)(*fog*)(*blowing snow*)]**.

Some weather maps display weather conditions at individual weather stations by the use of a station model. One such model is shown in the course website's User's Guide. Go to the course website and under Extras, click on User's Guide. Then, under the "Extras" heading, click on *Map Symbols*. Refer to the Surface Station Model and related explanations of map symbols to interpret the plotted weather data in the red box of the map to the right below:

7. Temperature: **[(*25*)(*28*)(*35*)]** °F.

8. Dewpoint: **[(*20*)(*25*)(*30*)]** °F.

9. Wind direction is shown by the "arrow" shaft drawn into the circle representing the station. North is to the top on the map and east is to the right. Wind is always named for the direction *from which* it blows. In the above depiction, the wind direction is generally from the **[(*northwest*) (*southwest*)(*south*)]**.

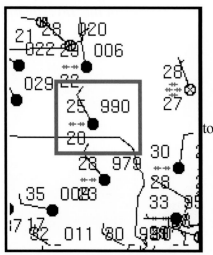

10. Wind speed is rounded off to the nearest 5 knots (1 knot equals 1.2 miles per hour) and symbolized by "feathers" drawn on the clockwise side of the wind-direction shaft. A full feather represents 10 knots and half feather indicates 5 knots. A pennant flag indicates 50 knots. A wind-direction shaft without feathers depicts a 1-to-2 knot wind and a circle drawn around the station circle signifies calm conditions. In this case, the reported wind speed is **[(*1-2*)(*5*)(*10*)(*50*)]** knots.

11. Air pressure (adjusted to sea level) is reported in the station model as a coded number to the nearest tenth of a millibar (mb). To decipher the plotted pressure value, first place a decimal point between the second and third numbers from the left. Then add either a "9" or "10" to the left so that the resulting number falls within the range of air pressures that could occur at sea level (almost always between 960 mb and 1050 mb). For example, a plotted value of 126 represents 1012.6 mb and 863 denotes 986.3 mb. The air pressure reported above is **[(*990.0*)(*999.0*)(*1099.0*)]** mb.

12. Sky cover is reported inside the station circle and is expressed as a number of eighths or other descriptors (scattered, broken, overcast, obscured). As examples, an empty circle indicates no clouds, and a half-shaded circle means four-eighths of the sky is cloud-covered. According to the User's Guide, Extras, *Weather Map Symbols*, the reported cloud cover is **[(*3/8*)(*7/8*)(*overcast*)]**.

13. Current weather is plotted at the "9 o'clock" position on the station model (to the left of the station circle) using a variety of symbols representing the particular weather conditions. According to the User's Guide, Extras, *Map Symbols*, the reported current weather is **[(*rain*)(*drizzle*)(*snow*)]**.

As directed by your course instructor, complete this investigation by either:

1. *Going to the Current Weather Studies link on the course website, or*
2. *Continuing to the Applications section for this investigation that immediately follows in this Investigations Manual.*

Investigation 2A: Applications

SURFACE WEATHER MAPS

A storm system traveling across the Southeastern states bringing heavy rain to the Gulf of Mexico coastal states and record-setting snowfalls to the Mid-Atlantic region will be examined in the Application. The weather conditions on Friday morning, 5 February, as the storm system took aim at the Atlantic coast area were given at specific stations in a coded station model.

14. **Figure 2** is the "U.S. - Data" map for Friday morning, 14Z 5 FEB 2010 (9 AM EST, 8 AM CST, 7 AM MST, 6 AM PST). The Figure 2 map shows selected weather data plotted about circles that represent the location of stations using the coded surface station model. Observe the station model for Atlanta, Georgia. The station model shows a temperature of **[(_42_)(_38_)(_29_)]** degrees F, and a dewpoint of 36 degrees F.

15. The winds at Atlanta were generally <u>from</u> the **[(_north_)(_east_)(_south_)(_west_)]** at about 20 knots (two long "feathers" on the shaft).

16. The coded pressure value was plotted as "_104_", meaning the actual atmospheric pressure corrected to sea level was **[(_10.4_)(_104.0_)(_1010.4_)(_1104.0_)]** mb.

17. The sky cover (amount of shading inside the station circle) indicated **[(_clear_)(_partly cloudy_)(_overcast_)]** conditions existed over Atlanta at map time.

18. Wind speeds across the map showed a variety of station model notations for different speeds. Charleston, SC had one long and one short feather for 15 knots while Tallahassee, in the Florida panhandle to the south, had one short feather for 5 knots. Albuquerque, New Mexico had a circle around the station circle to denote a wind speed reported as **[(_calm_)(_25 knots_)(_50 knots_)]**.

19. At the "9 o'clock" position alongside the station circle, a weather symbol may be plotted to show present weather conditions. Atlanta shows two dots that signify **[(_light rain_)(_light snow_)(_haze_)(_fog_)]** was occurring. Nashville, Tennessee also had two dots. Tallahassee had four dots designating heavy precipitation. For a listing of the frequently occurring weather symbols, see the *User's Guide* linked from the course website or go to *http://www.hpc.ncep.noaa.gov/html/stationplot.shtml*.

20. Other stations were reporting various significant weather conditions. The weather symbol at Jackson, Mississippi was two horizontal lines. This symbol indicated that Jackson was experiencing **[(_rain_)(_fog_)(_haze_)]**.

21. The figure "8" on its side at Charleston, SC, meant **[(_rain_)(_fog_)(_haze_)]** was present there.

22. Des Moines, Iowa was showing three "stars" as the symbol for moderate rates of [(***rain***) (***snow***)(***haze***)(***fog***)] at map time.

Figure 3 is the "Isobars, Fronts, Radar, & Data" map for Friday morning, 14Z 5 FEB 2010, the same time as the Figure 2 U.S. - Data map. This map additionally shows the position of high-pressure centers and low-pressure centers with fronts as analyzed by meteorologists at NOAA's National Centers for Environmental Prediction. Colored shadings further denote where the national network of weather radars had detected precipitation. Radar reflectivities are related to precipitation intensities according to the color scale along the left margin of the map area. (Generally, the greater the reflectivity the greater the intensity of precipitation.)

23. The coded station pressure reports were used to draw a computer analysis of the pressure pattern similar to the one you did in Investigation 1A, **Applications**. A center of relatively [(***low***)(***high***)] pressure was shown in central Alabama. This was the center of the storm system about to make history over central portions of the East Coast.

24. In the broad area of precipitation covering much of the eastern half of the country, the individual station sky cover symbols [(***are***)(***are not***)] mostly cloudy to overcast in the region.

25. The wind directions at the stations in the area about the **L** centered in Alabama were generally [(***counterclockwise and inward***)(***clockwise and outward***)], consistent with the hand-twist model of a Low.

26. One heavy line with triangles (blue on-screen) from southern Alabama southward into the Gulf of Mexico marked the position of a [(***cold***)(***warm***)(***stationary***)] front. This front was the leading edge of slightly cooler air that was advancing eastward and producing thunderstorm activity along the Gulf coast.

27. A short heavy line with half-circles (red) designated a [(***cold***)(***warm***)(***stationary***)] front that extended from the Low to the east and off-shore of South Carolina. The southeasterly flow noted at Charleston and Jacksonville, FL identified warm, humid air from the Atlantic riding up and over colder air ahead of the front in the coastal states. This flow of humid air led to the massive snowfalls over the next two days in the Mid-Atlantic area.

28. Along much of the northern tier of states a frontal system was shown with alternating blue triangles and red semicircles on opposite sides designating this portion of the frontal system to be a [(***cold***)(***occluded***)(***stationary***)] front.

Additionally, dashed orange lines on the map signify low-pressure troughs, extensions of lower pressure, sometimes marking the positions of dissipated or forming fronts.

For more practice on deciphering station models and map symbols, go to *http://profhorn.aos. wisc.edu/wxwise/AckermanKnox/chap1/decoding_surface.html*.

Examining a sequence of recent surface weather maps ending with the current map can show the movement of "weather makers" (high and low pressure centers and fronts) and the changes in atmospheric conditions at your location over time resulting from their movements. Practice looking for connections between weather changes depicted on the map sequence and predict local weather for the next half day or so.

You can also observe how station model conditions change as weather systems pass across a region by going to the *NWS Surface Analysis* link on the course website. The link takes you to the NWS Hydrometeorological Prediction Center site. Scrolling down allows you to "Create a Surface Analysis Loop" for North America or various regions. The **United States** choice, then clicking "Display Loop" will show an animation of the latest day's series of 3-hourly station models along with fronts and Highs and Lows.

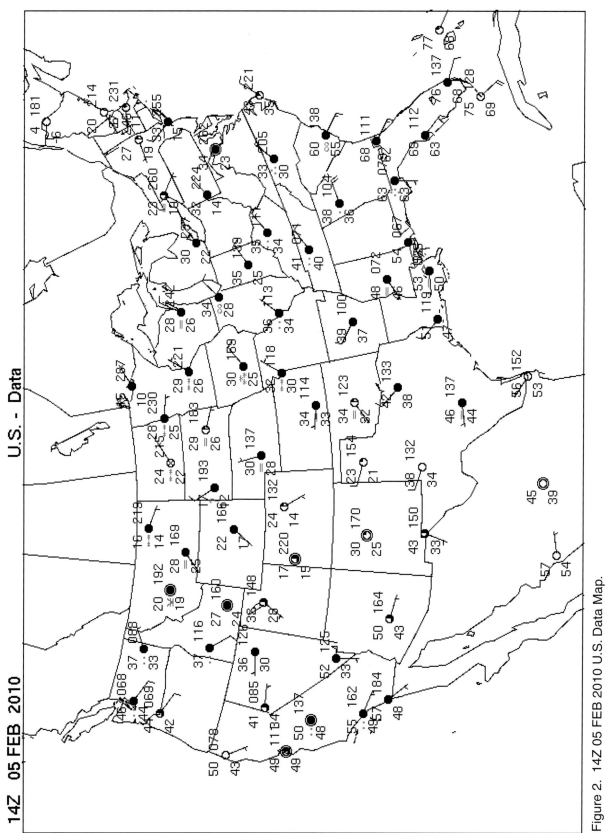

Figure 2. 14Z 05 FEB 2010 U.S. Data Map.

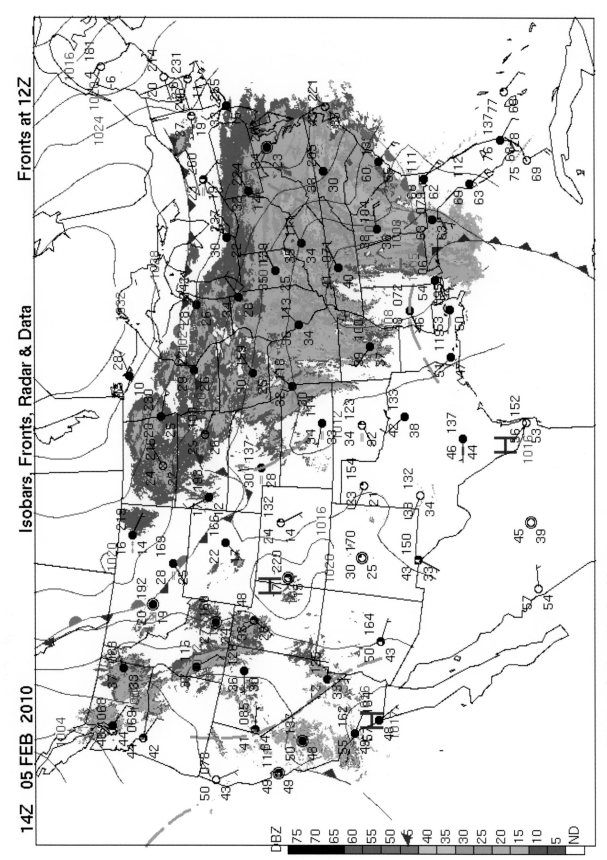

Figure 3. 14Z 05 FEB 2010 Surface Weather Map.

Investigation 2B:

THE ATMOSPHERE IN THE VERTICAL

Objectives:

The atmosphere has depth as well as horizontal dimensions. For a more complete understanding of weather, knowledge of atmospheric conditions in the vertical is necessary. Air, a highly compressible fluid held to the planet by gravity and squeezed under its own weight, thins rapidly with increasing altitude. The atmosphere is heated primarily from below, is almost always in motion, and contains a substance (water) that continually cycles through it while undergoing changes in phase.

After completing this investigation, you should be able to:

- Describe the vertical temperature of the atmosphere in the troposphere (the "weather" layer) and in the lower stratosphere.
- Compare the temperature profile specified by the U.S. Standard Atmosphere with actual soundings of the lower atmosphere.

Introduction:

Figure 1 shows the average vertical temperature profile of essentially the entire atmosphere as a function of the altitude above Earth's surface.

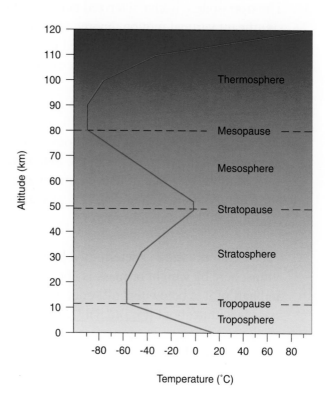

Figure 1.
Variation of average temperature with altitude
in the atmosphere.

Figure 2 is a Stüve diagram, a temperature/pressure graph of the lower portion of the atmosphere, used for plotting atmospheric conditions. Figure 2 focuses on the lowest 16 km of Figure 1.

1. **Plot the data points given below onto Figure 2 and connect adjacent points with solid straight lines.** Note that altitude is plotted along the right vertical axis increasing from bottom to top, and temperature is plotted along the horizontal axis increasing from left to right.

Altitude (km)	Temperature (°C)
16	– 56.5
11	– 56.5
0	+ 15.0

You have drawn the temperature profile of the lower portion of the "U.S. Standard Atmosphere." The Standard Atmosphere describes representative or average conditions of the atmosphere in the vertical. As seen in Figure 1, the temperature profile from the surface to 11 km depicts the lowest layer of the atmosphere, called the [(***troposphere***) (***stratosphere***)(***mesosphere***)(***thermosphere***)].

2. The lower portion of the [(***troposphere***)(***stratosphere***)(***mesosphere***)(***thermosphere***)] is evident immediately above 11 km in Figure 2 where temperatures remain steady with increasing altitude.

3. The troposphere is characterized generally by decreasing temperature with altitude, significant vertical motion, appreciable water vapor, and weather. According to the Standard Atmosphere data provided in item 1 above, the temperature within the troposphere decreases with altitude at the rate of about [(***4.5***)(***5.1***)(***6.5***)] C degrees per km.

4. Air pressure is plotted along the left vertical axis of the figure in millibars (mb), with pressure decreasing upward as it does in the atmosphere. Air pressure, which is very close to 1000 mb at sea level in the Standard Atmosphere, decreases most rapidly with altitude in the lowest part of the atmosphere. The Figure 2 diagram shows that an air pressure of 500 mb (about half that at sea level) occurs at an altitude of about [(***5.5***)(***8.3***) (***11.0***)(***16.0***)] km above sea level.

5. Because air pressure is determined by the weight of the overlying air, half of the atmosphere by weight or mass is above the altitude at which the air pressure is 500 mb and half of it is below that altitude. In other words, half of the atmosphere by weight or mass is within about [(***5.5***)(***8.3***)(***11.0***)(***16.0***)] km of sea level.

6. Other pressure levels can be found similarly. For example, 10% of the atmosphere is located <u>above</u> the altitude where the pressure is [(***100***)(***900***)] mb.

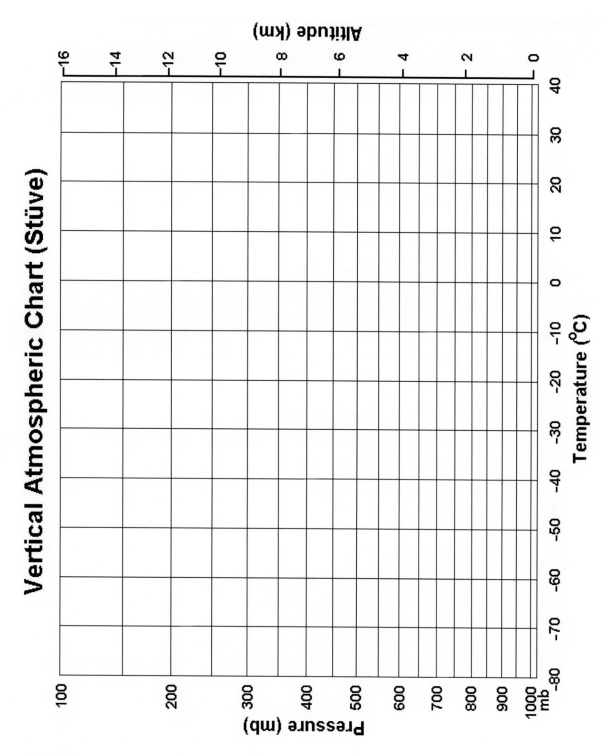

Figure 2.
Vertical atmospheric chart (Stüve diagram).

7. In other words, it can be seen in Figure 2 that at the altitude of approximately [(**5.5**)(**8.3**) (**11.0**)(**16.0**)] km above sea level, 10% of the atmosphere by weight is above and 90% is below.

As directed by your course instructor, complete this investigation by either:

1. *Going to the Current Weather Studies link on the course website, or*
2. *Continuing to the Applications section for this investigation that immediately follows in this Investigations Manual.*

Investigation 2B: Applications

THE ATMOSPHERE IN THE VERTICAL

The winter of 2009-10 was a record-breaking season for many of the cities of the eastern U.S. Several heavy snow storms crossed the metropolitan areas of the East. The storm system investigated in this Applications was a combination of two storms, one that crossed over the upper Midwest dumping on Chicago and another that moved up from the Southeast merging over Washington, DC. **Figure 3** is the surface weather map (Isobars, Fronts, Radar & Data) for 00Z 10 FEB 2010. The complex storm system is shown with one low-pressure center in north-central Ohio and two additional Lows along the northern Florida to South Carolina coasts.

8. Detroit, in southeastern Michigan, was shown with a plotted temperature at map time of [(*32*)(*26*)(*14*)] degrees F and a dewpoint of 23 degrees.

9. The present weather symbol (in the "9 o'clock" position) of two stars indicated that [(*rain*)(*snow*)(*fog*)] was occurring. Radar shadings showed that precipitation was spread in a broad arc from the northern Great Lakes along the East Coast to southern Florida.

In addition to this description of the weather at Earth's surface, conditions above the surface are also needed. Upper atmosphere data are collected by radiosondes at 00Z and 12Z each day from about 70 stations across the U.S. as part of a worldwide effort of sensing the three-dimensional atmosphere for weather analysis and forecasting. The following selected data were obtained by the rawinsonde (a radiosonde instrument also tracked for wind information) observation from Detroit, Michigan for 00Z 10 FEB 2010, the same time as the surface map. Plot these temperatures at their respective pressure levels over Detroit on the Stüve diagram of the prior portion of Investigation 2B (page 2B-5). The course website provides a link in the "Extras" section for an additional copy of the Stüve diagram if necessary. [*Note:* The pressure scale is <u>decreasing</u> upward along the left side.] Connect these points with a dashed line to distinguish from the Standard Atmosphere plot.

Pressure (mb)	Temperature (°C)	Altitude (m)
100	−55.1	15950
150	−47.9	13330
263	−55.7	9656
400	−38.9	6890
500	−28.5	5330
700	−12.1	2826
850	−7.5	1324
965 (surface)	−4.9	329

10. From the surface to about the 230-mb pressure level, temperatures in the atmosphere over Detroit on 10 February at 00Z were [(*__warmer than__*)(*__about the same as__*)(*__colder than__*)] than Standard Atmosphere conditions.

11. The tropopause is a defined boundary that separates the troposphere where temperatures generally decrease with increasing altitude, from the stratosphere where temperatures are steady (termed *isothermal*) or nearly so, or they increase with altitude (called a temperature *inversion*). Based on the temperature data you plotted, the tropopause above Detroit was located at a pressure level of [(*__500__*)(*__400__*)(*__263__*)] mb at 00Z on 10 February 2010.

12. The tropopause over Detroit at this time occurred at [(*__a lower__*)(*__the same__*)(*__a higher__*)] altitude than in the Standard Atmosphere. (Recall from the first part of this investigation, the Standard Atmosphere tropopause occurs at 11 km or 11,000 m.)

13. From the table of radiosonde data, the pressure of 500 mb occurred over Detroit at 00Z on 10 FEB 2010 at an altitude of [(*__5120__*)(*__5330__*)(*__5590__*)] m. Five hundred millibars is about one-half of the atmospheric pressure at sea level. Since air pressure is determined by the weight of the overlying air, this means one-half of the mass of Detroit's atmosphere is above this altitude and one-half below.

14. In the Standard Atmosphere a pressure of 500 mb occurs at an altitude of 5574 m (18,289 ft.). The altitude of the 500-mb level over Detroit at the time of the sounding was [(*__higher than__*)(*__the same as__*)(*__lower than__*)] that of the Standard Atmosphere. Therefore, the lower half of Detroit's atmosphere at the time of this observation is vertically more "contracted" from the surface to 500 mb which is consistent with the existence of colder air.

15. Our Stüve diagram is scaled to allow plotting of atmospheric data up to a pressure of 100 mb (about 16 km). If the atmospheric pressure at that level is 100 mb, about [(*__50%__*)(*__25%__*)(*__10%__*)] of the atmosphere remains above. We have not reached the top of the atmosphere!

Figure 4 is the plotted Stüve diagram for Detroit, Michigan (DTX) at 0000Z 10 FEB 2010 (labeled *100210/0000*) from the course website (**Upper Air**, "Stüves for Selected Cities"). This Stüve is plotted using all the data from the radiosonde observation. On the website, under the **Upper Air** section, "Upper Air Data - Text" provides the tabular listing of all data from observations that correspond to the latest plotted soundings. On the graph, the heavy plotted curve to the right represents temperatures. Winds at various levels are plotted to the right of the graph area using the convention of the surface model. For example, the winds at about 100 mb (highest arrow point) were <u>from</u> the west-southwest at about 55 knots. Additional lines on the Stüve diagram will be discussed in a later investigation.

16. On the Figure 4 Stüve diagram, note the temperature pattern from about 860 mb up to 850 mb and again from about 810 to 790 mb. Each of these layers is an example of [(*__an__*

**isothermal layer**)(_**a temperature inversion**_)(_**normally decreasing temperatures**_)]. These departures from the usual temperature profile ("wiggles") at lower levels are a frequent occurrence in the tropopause and show the complexity of atmospheric conditions. Knowledge of the finer structure of the temperature profile is often very important in interpreting atmospheric processes and motions above the surface.

The actual Detroit 00Z 10 February 2010 rawinsonde sounding reported atmospheric conditions to 12 mb (29,394 m) where the balloon burst. There was still some unsampled air above that! This cold atmospheric column was associated with the extensive storm system seen on the surface map in Figure 3 bringing a second paralyzing snowfall to the East Coast within a week.

For additional examples, look at the Stüve diagrams for Anchorage, Alaska and Hilo, Hawaii from the Upper Air section. Relative to the Standard Atmosphere, Alaska is typically cooler and Hawaii warmer. Below the table of cities on the Upper Air Stüves webpage or text data webpage, a link ("click here") provides access to upper air data and rawinsonde plots worldwide.

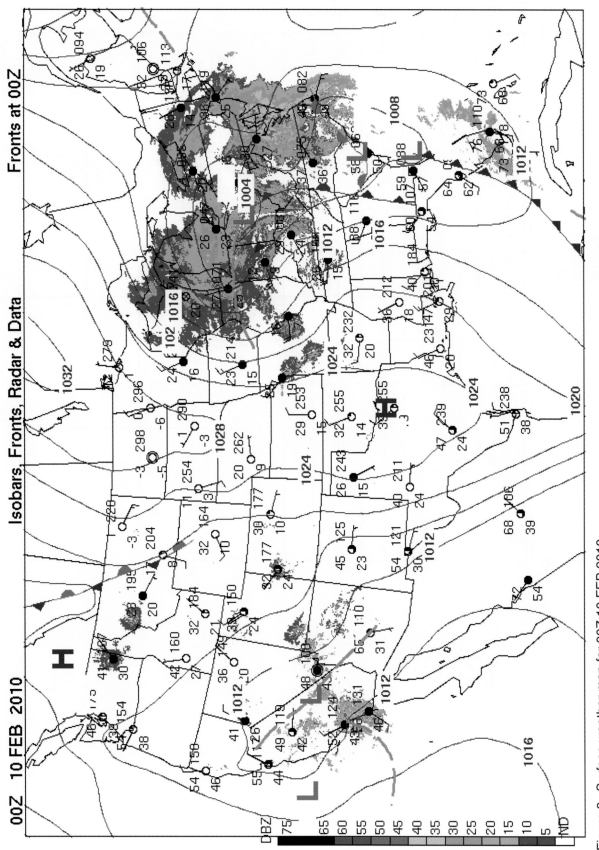

Figure 3. Surface weather map for 00Z 10 FEB 2010.

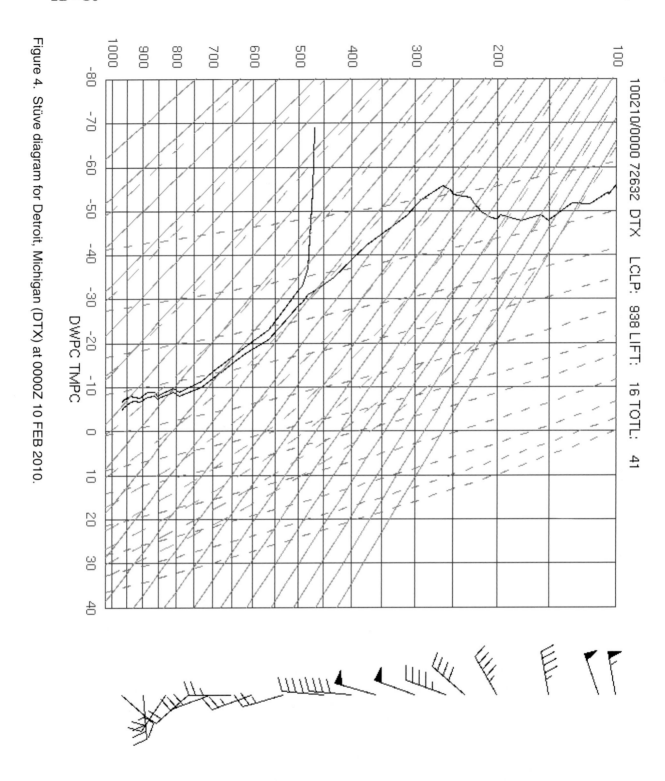

Figure 4. Stüve diagram for Detroit, Michigan (DTX) at 0000Z 10 FEB 2010.

Investigation

3A:

WEATHER SATELLITE IMAGERY

Objectives:

Orbiting satellites are platforms carrying sensors that make it possible for us to "look" down on the atmosphere and underlying Earth surfaces. It is obvious from the broad expanses seen from their vantage points that fair and stormy weather are somehow related. Clear areas and giant whirls of clouds blend one from another within the canopy of air that thinly envelops the planet. Over time, weather systems can be observed as they evolve, swirl, and voyage across Earth's surface. With the satellite views, areas showing signs of potential or actual hazardous weather conditions can be carefully monitored.

Satellite images are produced by sunlight that is reflected (and scattered) by the Earth-atmosphere system and by radiation that is emitted by that same system.

After completing this investigation, you should be able to:

- Distinguish among the different types of weather-satellite imagery and describe the information they can provide.
- Interpret probable atmospheric conditions from weather-satellite imagery.

Introduction:

The accompanying images in **Figure 1** were acquired simultaneously from sensors aboard an Earth geostationary satellite in orbit about 36,000 km (22,300 mi) above a spot on the equator at 75 degrees West Longitude. They were acquired on 22 March 2009 at 2345Z; about two days after the spring equinox. A geostationary satellite orbits toward the east at the same rate as the Earth rotates eastward so that the satellite appears to be hovering above the same place on Earth's surface.

1. Continental outlines are superimposed on the satellite images for orientation. A small "+" is marked in the center of each image approximately where the horizontal equator intersects the vertical 75°W longitude line. This marks the location of the sub-satellite point, that is, the spot on Earth's surface directly under the satellite. The sub-satellite point is located in [(***South***)(***North***)] America.

One satellite image was produced by reflected sunlight and the other by infrared radiation. **Keeping in mind that the Earth radiates infrared radiation continually (day and night), label the appropriate images as "visible" or "infrared".**

2. On the visible satellite image, the Sun's rays are from the general direction of [(***east***) (***west***)].

3. Because the Earth rotates eastward, local time at the sub-satellite point is near [(***sunrise***)(***sunset***)].

4. Using reflected visible light, the satellite sensor "sees" clouds and surface features as we do. In the visible image, the general appearance of all the clouds in the illuminated portion of the image is [(***white***)(***dark***)].

5. The expanse of clouds across parts of the North and South Pacific Oceans illustrates this point. Compared to the land and ocean, clouds have a [(***lower***)(***higher***)] albedo. (*Albedo* is the percentage of the sunlight striking a body that is reflected from it.)

6. The broad-scale organization of clouds provides clues as to the types and locations of various weather systems. On the visible satellite image, a large swirl of clouds characterized a large storm system that is evident over the [(***North Pacific Ocean***)(***mid-North American continent***)].

7. An important advantage of infrared imagery is that it can be used to observe the planet both day and night. The image produced by infrared radiation emitted by the Earth-atmosphere system demonstrates that there are clouds in [(***only the daylight portion***)(***both the daylight and night portions***)] of the Earth view shown.

8. In the infrared image, relatively warm land and sea surfaces appear dark, cooler low cloud tops are gray, and cold high cloud tops are shown as bright white. Therefore, the contrast in the appearance of blobs of clouds across South America and in an east-west band south of the equator versus those over the South Pacific Ocean west of the indentation of the South American coast indicates that the tops of the cloud blobs are at relatively [(***low***)(***high***)] levels.

9. In the infrared image, we can infer that the arcing swirl of clouds associated with the storm system mentioned earlier over North America has relatively [(***high***)(***low***)] tops.

As directed by your course instructor, complete this investigation by either:

   1. *Going to the Current Weather Studies link on the course website, or*
   2. *Continuing to the Applications section for this investigation that immediately follows in this Investigations Manual.*

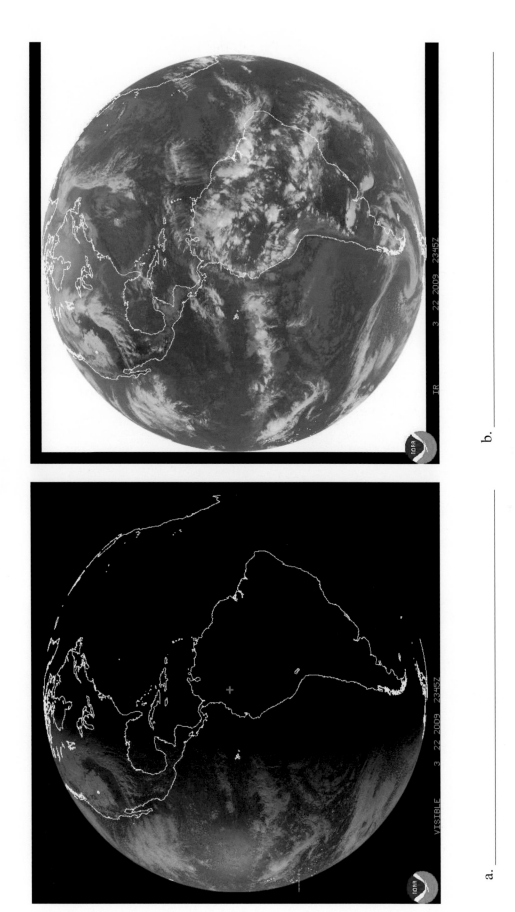

Figure 1.
Geostationary full-disk satellite images from 2345Z 22 March 2009.

Investigation 3A: Applications

WEATHER SATELLITE IMAGERY

At the time of the accompanying satellite images (Figures 2 and 3), it was well along through the winter season. Meteorological winter comprises the months of December, January and February. The astronomical winter runs from the winter solstice in December to the spring equinox [1732Z (1:32 PM EDT, etc.) 20 March 2010] when the daily periods of daylight and darkness are approximately equal. On the equinox they are essentially equal everywhere on Earth except at the poles (allowing for the few minutes of extended daylight due to atmospheric refraction). That means that on the spring equinox, Earth's Northern and Southern Hemispheres are equally illuminated. And the local sunrise and sunset positions on your horizon are due east and west, respectively, at the equinox.

Figure 2 is the *visible* satellite display for 0015Z on 15 FEB 2010 (6:15 PM Central Standard Time on 14 February, Valentine's Day), labeled across the upper margin "Visible Image" (listed "Visible - Latest" on the course website). The image is for the time when sunset was occurring along a line stretching from eastern Texas to western North Dakota on Sunday evening.

10. The visible image shows the cloud patterns in the western U.S. Major expanses of cloud cover the area from British Columbia Province in Canada to northern California and Nevada and also New Mexico. (The Olympic venues at Vancouver which was occurring at the time were still in daylight). Meteorologically, a cold front had moved into central Washington and Oregon bringing rain and snows to the higher elevations of the region. The dense clouds in the region with a band extending into the Pacific off northern California mark the location of the frontal system at this time. At the time of the visible satellite image, sunlight was reaching the U.S. from the general direction of [(***southeast***) (***southwest***)].

11. Cloud conditions, if any existed at this time, are not seen across approximately the eastern half of the coterminous U.S. because [(***the Sun had already set leaving darkness***) (***the satellite is not aimed to view the surface***)].

Note the short red marks located just beyond the upper and lower map boundaries. Draw a straight line across the satellite view connecting these two marks. The line you drew represents the "terminator" or line separating day and night at the time of Figure 2 (0015Z on 15 FEB 2010). Label this line "**terminator**". This can also be called the sunset line as it is progressing westward across Earth's surface as the planet rotates eastward. At the time sunset was occurring at Crosby, near the western North Dakota border near the U.S.-Canadian line at longitude 103.3 °W, latitude 48.9 °N, and Palacios, in eastern Texas along the Gulf Coast at longitude 96.2 °W, latitude 28.7 °N.

12. The western borders of much of Missouri and Arkansas as well as those of the Dakotas and Nebraska represent approximate north-south longitude lines. The terminator line you

drew is [(*parallel*)(*at a considerable angle*)] to the north-south border longitude lines. The terminator's orientation relative to north-south longitude lines changes throughout the year and will be discussed in this week's second activity. You can also note that the longitudes of the sunset cities listed are several degrees different.

13. **Figure 3** is the *infrared* satellite image (from "Infrared - Latest") for the same time (0015Z on 15 FEB 2010) as the visible image. This image shows much [(*less*)(*more*)] cloudiness in the eastern half of the country, particularly in a broad, curving arc from Illinois southwestward to Texas, and also across southern Florida as well as over the Atlantic, than was seen in the visible image. Infrared images are basically temperature maps of the surfaces "seen" by the satellite sensor.

14. Warm surfaces (land during most of the year and low clouds) appear relatively [(*bright white*)(*dark*)].

15. Cold surfaces such as high cloud tops, emitting little infrared radiation appear [(*bright white*)(*dark*)]. Surfaces with intermediate temperatures appear in gray shadings.

16. At 0015Z 15 FEB 2010 several areas of rain and snowshowers were spread across the U.S. One particularly notable area of thunderstorms stretched from west-central Mississippi across Louisiana to the east Texas Gulf Coast. The relative brightness of those "bumpy" cloud tops is evidence they are generally [(*"warm"*)(*"cold"*)].

On the Figure 2 visible satellite image you might also be able to see the "bumpy" texture of clouds over the western U.S. At least some of this effect comes from the shadows produced from towering cloud tops on lower clouds to the east due to the low sun angle.

17. Back on the infrared satellite Figure 3, bright cloud tops associated with the upper atmosphere subtropical jetstream over southern Florida were generally at [(*low*)(*high*)] altitudes.

18. Comparatively, the white area of scattered clouds seen in the visible image located offshore from southern California and the Baja California peninsula of Mexico were not distinguishable in the shadings along the coastal Pacific in the infrared image. These cloud tops offshore of California and Mexico were relatively [(*"warmer"*) (*"colder"*)] than the thunderstorm tops and curved band over southern Florida. These offshore clouds over the Pacific were not evidenced in the infrared image because the coastal clouds were at lower altitudes, with temperatures of their tops near those of the nearby water surfaces.

19. If you wished to create a complete 24-hour time-lapse of the cloud patterns across the U.S. such as those routinely seen on television using the satellite images from each hour, you should choose [(*visible*)(*infrared*)] images because the other type of images would appear black during nighttime hours.

Visible Image   0015Z 15 FEB 2010

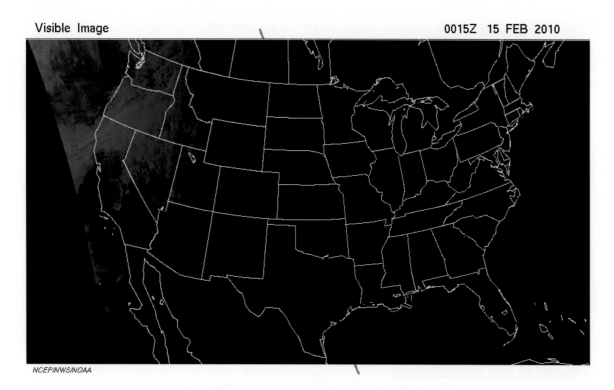

NCEP/NWS/NOAA

Figure 2. Visible satellite image at 0015Z 15 FEB 2010.

Infrared Image   0015Z 15 FEB 2010

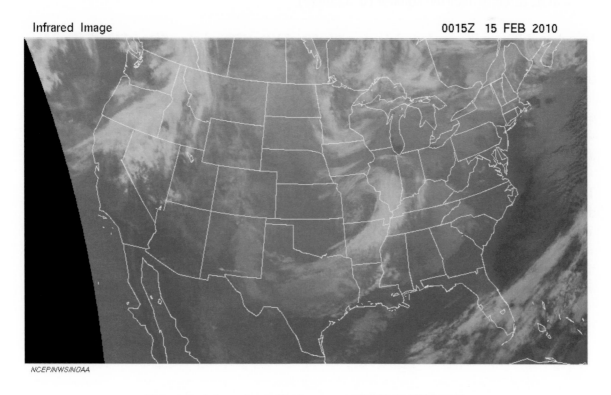

NCEP/NWS/NOAA

Figure 3. Infrared satellite image at 0015Z 15 FEB 2010.

Figure 4 is an image from the U.S. Naval Observatory web site, *http://aa.usno.navy.mil /data/docs/earthview.php*, at 6:15 PM CST, the same time as the satellite images. The depiction in Figure 4 is from a model showing the positions of the sunrise and sunset terminators and the portions of the entire Earth's surface in sunlight and darkness at that time. Mark the approximate positions of Crosby, ND and Palacios, TX from Figure 2 on the global map display in Figure 4.

20. The Figure 4 sunset terminator line across North America [(***did***)(***did not***)] closely coincide with the sunset line you drew on Image 1 where sunsets were 6:15 CST.

21. On the Figure 4 map, dotted lines represent horizontal parallels of latitude at 30° intervals with the equator (0°) in the middle and vertical meridians of longitude also at 30° intervals with the 0° Prime Meridian at either edge and the 180° International Date Line in the middle. The lengths of the lighted portions of horizontal lines (latitudes) at comparable distances north and south of the Equator, are proportional to the lengths of daylight at those latitudes (see for example 60° N and S latitudes). Their lengths infer that at this time of year (15 February) prior to the spring equinox, the daylight periods are [(***longer in the Northern Hemisphere***)(***longer in the Southern Hemisphere***) (***equal in both hemispheres***)].

22. The label beneath the Figure 4 map shows that at the time of the image, 0015Z 15 February 2010, the Sun was overhead at noon at 179.8° E longitude (essentially on the International Dateline) and [(***23.5***)(***0.7***)(***12.8***)] °S latitude. On the equinox, the noon Sun will be directly over the equator (0° latitude).

Based on different Figure 4 model views, you can determine what portion of Earth would be illuminated on the equinox or the summer solstice (or any other day of the year). Go to the Naval Observatory website listed above and enter the Year, Month, Day and time of day in Hours and Minutes Universal (Z) time. Finally select a Grayscale or Color view. Lastly, click **Show Earth Views**. In addition to the Mercator map view of the illuminated Earth at that time, there are polar views showing how the North and South polar regions are lit and also the view of the globe as seen from either the Sun or the moon.

The Naval Observatory website allows you to investigate whether or not the length of daylight changes on the equator from day to day. Check day and night segments in Figure 4 and then check day and night map views on the Naval Observatory website for the summer solstice on 21 June 2010. Also, sunrise and sunset times, along with other astronomical information, can be obtained from another U. S. Naval Observatory site, *http://www.usno.navy.mil/USNO/astronomical-applications/data-services/rs-one-day-us*.

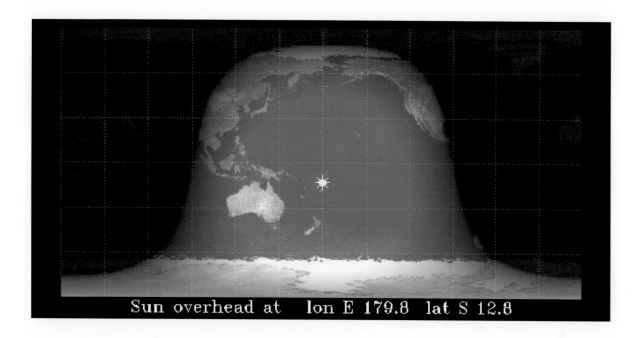

Sun overhead at   lon E 179.8   lat S 12.8

Figure 4. Model depiction of portions of Earth's surface in sunlight and darkness at 0015Z 15 FEB 2010. Subsolar position at image time is denoted by rayed spot.

SUNLIGHT THROUGHOUT THE YEAR

Objectives:

Ultimately, all weather and climate begins with the Sun. That is because solar radiation is the only significant source of energy that determines conditions at and above the Earth's surface.

The average rate at which solar radiation is received outside Earth's atmosphere on a surface oriented perpendicular to the Sun's rays is about 2 calories per square centimeter per minute (1370 W m^{-2}). The amount of solar radiation that actually reaches Earth's surface at any particular location is quite different.

The nearly-spherical Earth, rotating once a day on an axis inclined to the plane of its orbit, presents a constantly changing face to the Sun. Wherever there is daylight, the daily path of the Sun through the local sky changes through the course of a year. Everywhere on Earth, except at the equator, there is variation in the daily number of hours of daylight through the year. In addition, the atmosphere absorbs and scatters the solar radiation passing through it. Clouds, especially, can block much of the incoming radiation.

The purpose of this investigation is to consider the variability of sunlight received at different latitudes over the period of a year.

After completing this investigation, you should be able to:

- Describe the variation of solar radiation received at equatorial, midlatitude, and polar locations over the period of a year.
- Estimate and compare the amounts of sunlight received at equatorial, mid-latitude, and polar locations during the different seasons of the year.

Introduction:

Examine the accompanying graph of **Figure 1**. Data points plotted on the graph represent monthly averages of measurements of actual solar radiation received daily on a horizontal plane at Earth's surface at near equatorial (Singapore), midlatitude (Brockport, NY), and polar (South Pole, Antarctica) locations. On Figure 1, month of the year is plotted along the horizontal axis and average daily incident radiant energy in calories per square centimeter per day is plotted vertically. On the horizontal axis, the longer marks represent the first day of each month and the shorter marks represent mid-month.

Construct an annual solar radiation curve for <u>each</u> of the three locations. Do this by drawing a smooth curved line connecting the radiation values already plotted for each location (see map legend to identify plotted symbols). Note that at the South Pole (90 degrees S latitude) the Sun rises on or about September 23 and sets on or about March 21.

Draw each curve to the ends of the plotted values. December values are plotted twice to more clearly illustrate the annually repeating radiation cycles.

1. According to the curves you have drawn, the average daily solar radiation varies the <u>least</u> over the period of a year at the [(***equatorial***)(***midlatitude***)(***polar***)] location.

2. The variation in average daily solar radiation that does occur at the latitude identified in item 1 is primarily due to changes in the daily [(***period of sunlight***)(***path of sunlight through the atmosphere***)]. (Refer to Figures 2 through 6 when answering this and the following questions 3 through 7.)

3. The pattern of sunlight received at the equator over the course of a year indicates that the seasonal contrast there is [(***similar to***)(***much less than***)] the seasonal contrast experienced in midlatitudes.

4. The graph shows that there is a six-month period during which there is no sunlight at the [(***equatorial***)(***midlatitude***)(***polar***)] location.

5. According to the graph, there are months when both the midlatitude and polar locations receive more solar radiation than the equator. For both midlatitude and polar locations, the major factor that causes this difference is the greater [(***local noon solar altitude***)(***length of daylight each day***)].

6. Comparison of the three annual radiation curves indicates that the annual range (the difference between the curve's maximum and minimum) of solar radiation received daily [(***increases***)(***decreases***)] as latitude increases.

7. Based on how solar radiation received varies with latitude, it can be inferred that the seasonal temperature contrast [(***increases***)(***decreases***)] as latitude increases.

Mark the equinoxes and solstices on Figure 1 by drawing vertical lines at approximately March 21, June 22, September 23, and December 22. On the equinoxes the noon sun is at some point directly above the equator, whereas on the solstices the noon sun is at some point directly above 23.5 degrees N or 23.5 degrees S latitude. **Label the intervals between the lines as the <u>Northern</u> <u>Hemisphere's</u> winter, spring, summer, and fall seasons.**

8. Two maxima and two minima would appear in the annual solar radiation curve for the equatorial location if the sky were always clear. As seen from the actual Singapore data, it is evident that the maxima occur near the [(***solstices***)(***equinoxes***)].

9. Because the rate of radiation received at Earth's surface is plotted over time in Figure 1, the area enclosed under the curve in each of the seasonal segments is directly proportional to the total solar radiation received during that season. According to the "seasonal" areas under each curve, the [(***equatorial***)(***midlatitude***)(***polar***)] location receives more nearly the same total amount of solar radiation every season of the year.

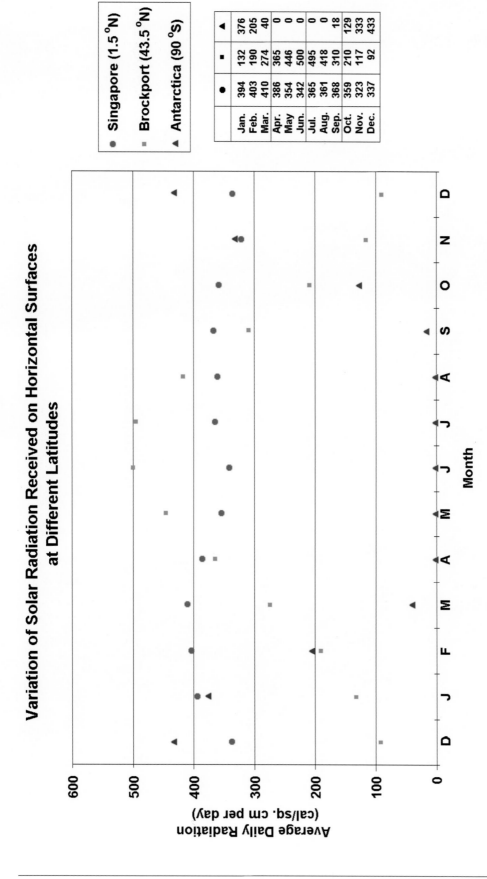

	●	■	◄
Jan.	394	132	376
Feb.	403	190	205
Mar.	410	274	40
Apr.	386	365	0
May	354	446	0
Jun.	342	500	0
Jul.	365	495	0
Aug.	361	418	0
Sep.	368	310	18
Oct.	359	210	129
Nov.	323	117	333
Dec.	337	92	433

● Singapore (1.5 °N)

■ Brockport (43.5 °N)

◄ Antarctica (90 °S)

Figure 1.
Variation of solar radiation received on horizontal surfaces at different latitudes.

10. At the midlatitude location, the seasons of [(*summer and fall*)(*fall and spring*) (*spring and summer*)] receive the <u>most</u> solar radiation.

11. For that same midlatitude location, the seasons of [(*fall and spring*)(*winter and fall*) (*fall and summer*)] receive the <u>least</u> solar radiation.

12. During spring and summer in the Northern Hemisphere, the South Pole receives [(*its maximum*)(*zero*)] solar radiation.

13. These are the Southern Hemisphere's seasons of [(*summer and fall*)(*fall and spring*) (*fall and winter*)].

As directed by your course instructor, complete this investigation by either:

1. *Going to the Current Weather Studies link on the course website, or*
2. *Continuing to the Applications section for this investigation that immediately follows in this Investigations Manual.*

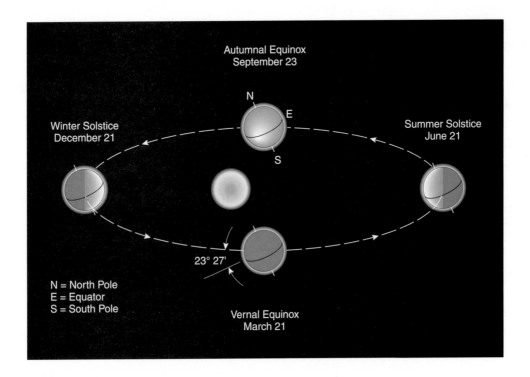

Figure 2. Earth's orbital relationship to the Sun on the solstices and equinoxes.

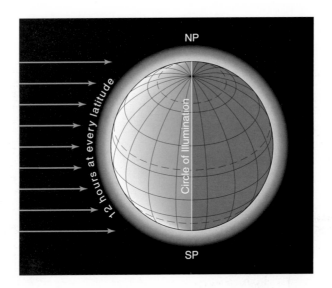

Figure 3. Solar radiation received at Earth on the equinox.

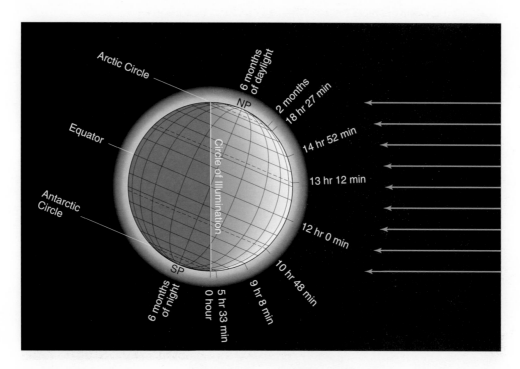

Figure 4. Solar radiation received at Earth on Northern Hemisphere's summer solstice.

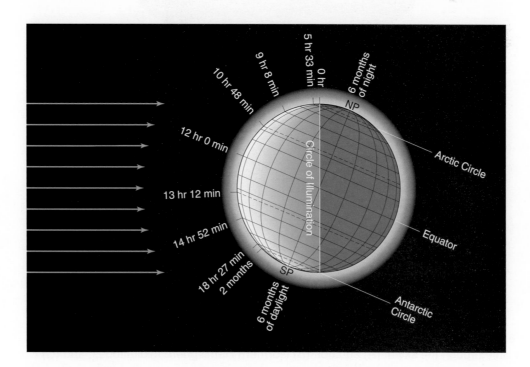

Figure 5. Solar radiation received at Earth on Northern Hemisphere's winter solstice.

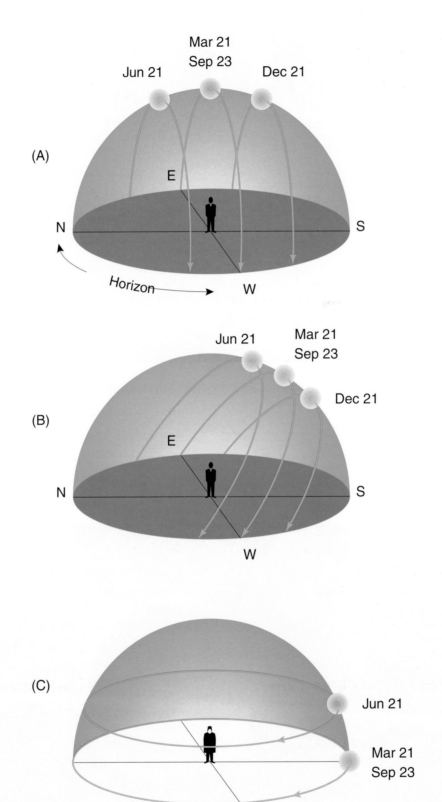

Figure 6.
Path of Sun through the sky at (A) equator, (B) Northern Hemisphere midlatitudes, and (C) North Pole.

Investigation 3B: Applications

SUNLIGHT THROUGHOUT THE YEAR

Examine the three visible satellite images in **Figures 7**, **8**, and **9**. These are actual images which were obtained on or near the first days of the Northern Hemisphere's fall, winter and summer seasons. Next, examine the small drawing to the right of each Earth image. The drawing shows the relative positions of Earth, the satellite, and rays of sunlight at the time each image was recorded. (In the small drawing, the view is from above Earth's Northern Hemisphere.) If you were located on the satellite, you would have seen the same view of Earth as shown in each accompanying satellite image.

The Earth images in Figures 7 to 9 were acquired when sunset was occurring at the point on the equator directly below the viewing satellite (in the center of the Earth's disk). Sunset was occurring along the dashed line passing through the sub-satellite point. The arrows to the left in each image represent incoming rays of sunlight at different latitudes. When answering the following questions, ignore the effects of Earth's atmosphere on the Sun's rays.

14. Look at Figure 7, the 23 September image. Note that the Earth's axis is perpendicular to the Sun's rays, so the sunset line and Earth's axis line up together. Also note that each latitude line, including the equator, is half in sunlight and half in darkness. Because the Earth rotates once in 24 hours, the period of daylight is **[(_0_)(_6_)(_12_)(_18_)(_24_)]** hours everywhere except right at the poles.

15. Now look at Figure 8, the 21 December satellite image. On the Northern Hemisphere's winter solstice, the Earth's North Pole reaches its maximum tilt away from the Sun for the year. Consequently, poleward from the Arctic circle, the daily period of daylight is **[(_0_)(_6_)(_12_)(_18_)(_24_)]** hours.

16. Examine the Northern Hemisphere latitude lines in the 21 December satellite image and compare how much of each line is in sunlight with the amount that is in darkness. The comparison shows that at all latitudes in the Northern Hemisphere of the rotating Earth, the daily period of daylight is **[(_greater than_)(_equal to_)(_less than_)]** the daily period of darkness.

17. Poleward from the Antarctic Circle on 21 December, the daily period of daylight is **[(_0_) (_6_)(_12_)(_18_)(_24_)]** hours.

18. Now look at Figure 9, the 21 June satellite image. This shows that on the Northern Hemisphere's summer solstice, Earth's North Pole attains it maximum tilt towards the Sun for the year. Consequently, poleward from the Arctic Circle, the period of daylight is **[(_0_)(_6_)(_12_)(_18_)(_24_)]** hours.

19. Poleward from the Antarctic Circle on 21 June, the period of daylight is [(_0_)(_6_)(_12_)(_18_) (_24_)] hours.

Along with these variations in the length of daylight at various latitudes as shown in the satellite views, the intensity of incoming sunlight varies with the angle of incidence of the Sun's rays striking Earth's surface. (A latitude line on Earth's disk and Sun's rays could also be added at your latitude.) Thus, the solar energy received at a location over the course of the year depends on the varying **solar altitude** (angle of the Sun above the horizon) and the **period of daylight** at that location.

20. **Return to Investigation 3A's Figure 1a.** This view is a satellite image for essentially the first day of the Northern Hemisphere's spring season. Approximate the position of the terminator. Its orientation (the sunset line in this view) most nearly looks like the orientation of the terminator line in the image for the first day of [(_winter_)(_fall_)(_summer_)] that you have just examined in this activity. This is because both are essentially equinox views.

Suggestions for Further Activities: As time progresses through the seasons, call up visible satellite images near times of local sunrise or sunset every week or so, and observe changes in the orientation of the terminator. For example, the Investigation 3A terminator line that you drew on Figure 2 in Investigation 3A was oriented slightly southeast – northwest (on 0015Z 15 February 2009). By the spring equinox, the terminator became south-north in orientation. Also, relate these satellite views to the path of the Sun through your local sky and the length of daylight at your location. And, note the general trend of temperatures relative to these changing conditions. Finally, you might call up *http://www.time.gov* to keep track of the changing length of daylight at various locations on the Earth.

Full disk satellite views like those of Figure 7 through 9 can be obtained from the course website under the **Satellite** section by clicking on "GOES Satellite Server", then selecting "GOES Full Disk" from the left side menu. Then click on one of the full disk "VIS"s or the images themselves to view an enlarged visible display. The visible image can also be compared with the infrared image for the same time. Check on these visible images for sunrise (near 12Z) or sunset (near 00Z) times to compare to this Investigation.

To estimate amounts of solar radiation received at your location for the various months of the year similar to that given in Figure 1 of Investigation 3B, you can call up *http://rredc. nrel.gov/solar/old_data/nsrdb/redbook/atlas/.* Because this site is designed to display the energy from solar collectors, you need to choose the options of: (a) type of data, such as "average", (b) month of year or annual, and (c) collector orientation - "horizontal flat-plate". Then click on "View The Map". Finally, the energy units of kilowatt hours per square meter per day estimated from the map need to be multiplied by 86.04 to obtain calories per square centimeter per day as used in this Investigation.

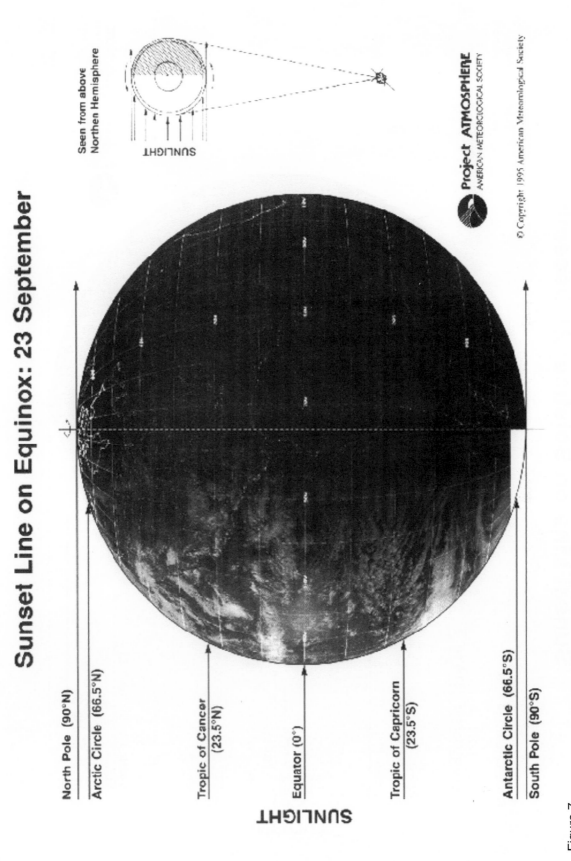

Sunset Line on Equinox: 23 September

Seen from above
Northen Hemisphere

SUNLIGHT

Project **ATMOSPHERE**
AMERICAN METEOROLOGICAL SOCIETY

© Copyright 1995 American Meteorological Society

North Pole (90°N)

Arctic Circle (66.5°N)

Tropic of Cancer
(23.5°N)

Equator (0°)

Tropic of Capricorn
(23.5°S)

Antarctic Circle (66.5°S)

South Pole (90°S)

SUNLIGHT

Figure 7.
Visible satellite image for 23 September.

Sunset Line on Solstice: 21 December

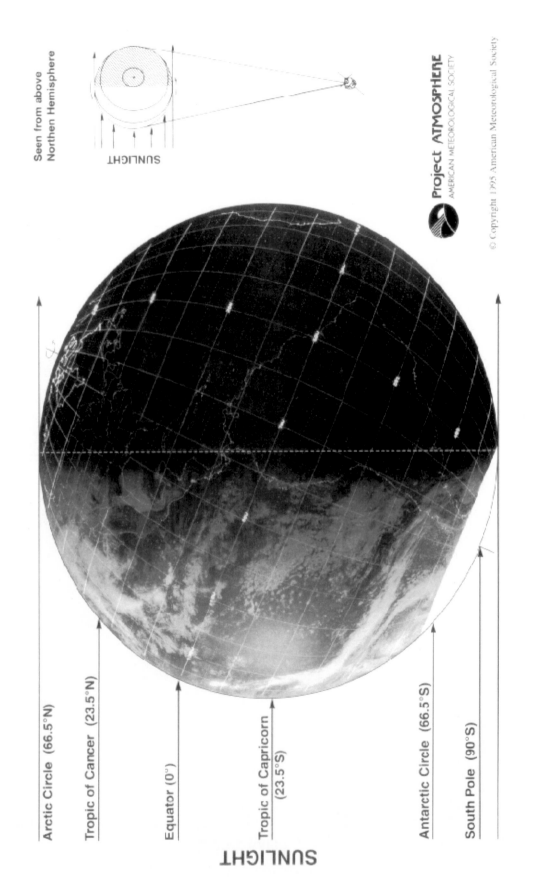

Seen from above
Northen Hemisphere

SUNLIGHT

Arctic Circle (66.5°N)

Tropic of Cancer (23.5°N)

Equator (0°)

Tropic of Capricorn (23.5°S)

Antarctic Circle (66.5°S)

South Pole (90°S)

SUNLIGHT

Project ATMOSPHERE
AMERICAN METEOROLOGICAL SOCIETY

© Copyright 1995 American Meteorological Society

Figure 8.
Visible satellite image for 21 December.

Sunset Line on Solstice: 21 June

Seen from above
Northen Hemisphere

SUNLIGHT

SUNLIGHT

North Pole (90°N)

Arctic Circle (66.5°N)

Tropic of Cancer (23.5°N)

Equator (0°)

Tropic of Capricorn (23.5°S)

Antarctic Circle (66.5°S)

Project ATMOSPHERE
AMERICAN METEOROLOGICAL SOCIETY

© Copyright 1995 American Meteorological Society

Figure 9.
Visible satellite image for 21 June.

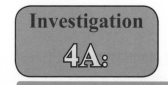
TEMPERATURE AND AIR MASS ADVECTION

Objectives:

The Earth-atmosphere system is heated unevenly by solar radiation. Low latitudes receive more energy from the Sun than they lose directly to space as outgoing infrared radiation. Averaged over a year, high latitudes experience more outgoing than incoming radiation. These energy excesses and deficits are balanced by movements (called advection) of heat energy poleward by migrating air masses, storm circulations, and ocean currents. The transport of air from a region of relatively high temperatures to a region of relatively low temperatures by air in motion (the wind) is referred to as *warm air advection*. Conversely, the transport of air from a region of relatively low temperatures to a region of relatively high temperatures by the wind is called *cold air advection*. Identification of areas of warm and cold air advection requires, in addition to information about winds, determination of the air temperature pattern made possible through the drawing of isotherms.

After completing this investigation, you should be able to:

- Draw lines of equal temperature (isotherms) to reveal the pattern of air temperatures across the nation at map time.
- Locate regions on a weather map where cold and warm air advection is likely to be occurring.
- Relate warm and cold air advection patterns to circulations of weather systems.

Introduction:

Temperature patterns are found on weather maps by drawing lines representing specific temperatures. These lines are called *isotherms* because every point on the same line has the same temperature value. The skills required to draw isotherms are much the same as those needed to draw isobars.

Tips on Drawing Isotherms:

a. Always draw an isotherm so that temperatures higher than its value are consistently to one side and lower temperatures are to the other side.

b. Assume a steady temperature change between neighboring stations when positioning isotherms; that is, use interpolation to place isotherms.

c. Adjacent isotherms tend to look alike. The isotherm you are drawing will often align in a general way with the curves of its neighbor because changes in air temperature from place to place are usually (but not always) gradual.

    d. Continue drawing an isotherm until it reaches the boundary of plotted data or "closes" within the data field by making its way to its other end and completing a loop.

    e. Isotherms can never be open ended within a data field and they never fork, touch, or cross one another.

    f. Isotherms cannot be skipped if their values fall within the range of temperatures reported on the map. Isotherms must always appear in sequence; for example, when the isothermal interval is 10 degrees, there must be a 50 °F isotherm between the 40 °F and the 60 °F isotherms.

    g. Always label isotherms.

The **Figure 1** map segment shows temperatures in degrees Fahrenheit (°F) at various weather stations. Consider each temperature value to be located at the center of the plotted number. The 70- and 60-°F isotherms have been drawn and labeled. **Complete the 50-°F isotherm. Then, draw the 40- and 30-°F isotherms. Be sure to label each isotherm at both ends.**

1. Isotherms are drawn at regular intervals; on this map, the interval between successive isotherms is **[(_5_)(_10_)(_20_)(_30_)]** Fahrenheit degrees.

2. Your temperature analysis reveals a pattern generally showing a decrease in temperature from the **[(_northeast to southwest_)(_northwest to southeast_)(_southeast to northwest_)]**.

3. In **Figure 2**, a surface weather map of the contiguous U.S. shows a simplified view of a storm system in the eastern part of the country. Local wind data are on display for several stations and three isotherms have been drawn on the map. The isotherm values (from lowest to highest) are **[(_50, 60_)(_60, 70_)(_50, 60, 70_)]** °F.

4. Hence, the interval between isotherms on the Figure 2 map is **[(_5_)(_10_)(_20_)(_30_)]** Fahrenheit degrees.

Warm air advection occurs where the wind blows across isotherms from the higher (warmer) values to the lower (colder) values. That is, warmer air is being transported towards a location by the horizontal winds. Based on this factor alone, one would expect temperatures at that location to be rising. *Cold air advection* occurs where the wind blows across the isotherms from the lower (colder) values to the higher (warmer) values. Then, colder air being transported by wind to the location produces falling temperatures.

Based on wind directions and the isotherm pattern on the Figure 2 map, determine the type of air advection (warm or cold) that would be occurring at each station.

5. Warm air advection was occurring at Station **[(_A_)(_B_)(_C_)]**.

6. Cold air advection was occurring at Station **[(_B_)(_C_)(_D_)]**.

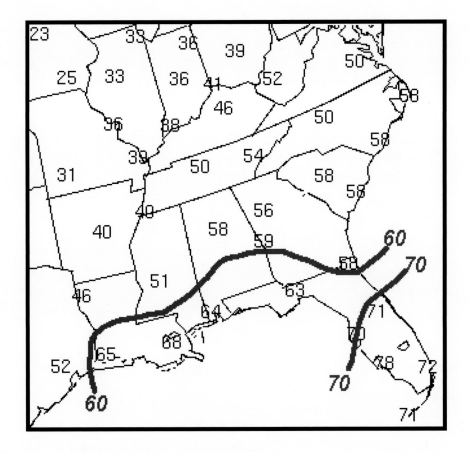

Figure 1.
Southeastern U.S. map segment of temperatures.

Surface Weather Map Showing Storm System in the Eastern United States

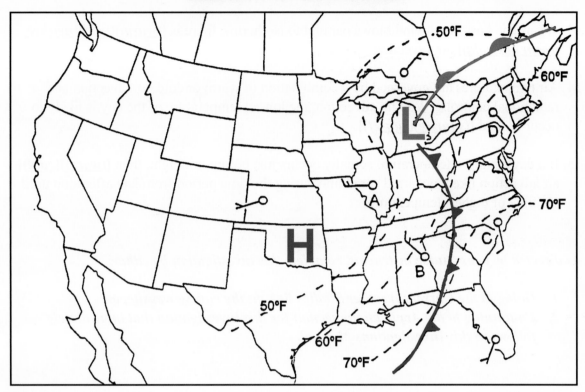

Figure 2.
Simplified surface weather map.

7. Generalizing from this map depiction of wind and temperature patterns associated with weather systems, areas southeast of Lows can be expected to have [(**_warm_**)(**_cold_**)] air advection.

8. Meanwhile, areas to the west and southwest of Lows can be expected to have [(**_warm_**)(**_cold_**)] air advection.

9. Areas to the east of Highs would be expected to have [(**_warm_**) (**_cold_**)] air advection.

10. And while not shown here but relying on the hand-twist model of a High, areas to the west of Highs should have [(**_warm_**)(**_cold_**)] air advection.

On actual weather maps, the patterns of isotherms and winds vary greatly. The intensity of warm and cold air advection will depend on the wind speeds, the angle at which the wind crosses the isotherms, and the closeness of neighboring isotherms. In general, the faster the wind, the more perpendicular the angle, and the closer the isotherms, the stronger the advection.

11. Not surprisingly, behind cold fronts one can expect [(***warm***)(***cold***)] air advection.

12. And behind warm fronts one can expect [(***warm***)(***cold***)] air advection.

13. Where the horizontal wind blows parallel to isotherms, there is [(***warm***)(***cold***)(***neither warm nor cold***)] air advection.

14. Air temperatures are governed by a combination of warm or cold air advection and radiational controls. With no advection, the lowest temperature of the day is likely to occur around [(***sunrise***)(***sunset***)].

15. If a day's highest temperature actually occurs just before midnight, then [(***warm***)(***cold***)] air advection likely occurred sometime during the time period from late afternoon until the time of highest temperature.

As directed by your course instructor, complete this investigation by either:

1. *Going to the Current Weather Studies link on the course website, or*
2. *Continuing to the Applications section for this investigation that immediately follows in this Investigations Manual.*

Investigation 4A: Applications

TEMPERATURE AND AIR MASS ADVECTION

On 22 February 2010, a complex storm system and its accompanying precipitation areas were crossing the eastern portion of the country. Significant snows were falling to the north of the track of the advancing low pressure center while rain and thunderstorms were occurring to the south of the passing storm's track. Those movements also brought relatively rapid changes of temperature for the locations over which they passed.

Figure 3 is the map of plotted station models with reported surface weather conditions (Isotherms, Fronts, & Data) for 11Z Monday, 22 FEB 2010 (6 AM EST, etc.). *Note that on this map, the isopleths are* **isotherms** *(not isobars).* The H and L centers and frontal positions are shown as of 09Z on the map.

The eastern U.S. storm system was primarily defined by a low-pressure center (**L**) and its accompanying fronts. The low center was shown located at the juncture of the Illinois, Indiana and Kentucky borders. A relatively warm air mass was located to the southeast of this storm center while another colder air mass was found to the storm's northwest. Centers of both of these air masses were marked by **H**s on the map.

16. From the storm's low-pressure center, a short **[(*warm*)(*cold*)(*stationary*)]** front was shown by a bold red line with half circles extending eastward. The frontal system then continued on to the East Coast as a stationary front.

17. From the Low center, a **[(*warm*)(*cold*)(*stationary*)]** front stretched to the south and southwest reaching the southern tip of Texas.

18. The winds at stations in the several-state area about the **L** displayed a circulation that was generally **[(*clockwise*)(*counterclockwise*)]** and inward. This circulation was bringing cold air southward from northern Plains states to the central portion of the country and warm air northward from the Southeast to the Mid-Atlantic States.

19. St. Louis, Missouri and Oklahoma City, OK were located near the **[(*20*)(*30*)(*40*)]** degree Fahrenheit isotherm and had temperatures within a few degrees of the isotherm's value. The isotherm label was located on the line in northern New Mexico. (The neighboring isotherms were labeled in western Kansas and northeastern Texas, respectively.) Highlight this isotherm from its position at the western edge of Lake Michigan to the Texas-New Mexico border.

20. St. Louis had winds plotted as 10 knots from the **[(*northwest*)(*northeast*)(*southwest*)(*southeast*)]** at map time. Oklahoma City's winds were from the north at 15 knots.

21. Generally, the winds at both stations exhibited flows that were directed across the

isotherm at large angles. St. Louis' winds were nearly perpendicular to the isotherm line. These wind directions were from [(**_higher toward lower_**)(**_lower toward higher_**)] temperature regions. Stations from Minnesota and Nebraska southward to the frontal system generally displayed this temperature-wind flow directional pattern.

22. Therefore, the wind flows across the isotherm at St. Louis and Oklahoma City demonstrated that [(**_cold_**)(**_warm_**)] air advection was occurring over this region.

23. This advection pattern was "behind" (to the west and north of) the advancing [(**_cold_**)(**_warm_**)] front.

24. One would conclude then that air motions and temperature patterns following a cold front would produce [(**_cold_**)(**_warm_**)] air advection.

25. By contrast, Lexington, Kentucky and Greensboro, North Carolina, had winds that were generally from a southerly direction. This flow was [(**_behind_**)(**_ahead of_**)] the warm front.

26. Greensboro was along the looping [(**_30_**)(**_40_**)(**_50_**)] degree Fahrenheit isotherm (labeled in southern West Virginia). Highlight this isotherm from the Low center to the Atlantic coast.

27. The isotherm and the direction of winds at Lexington and Greensboro show that [(**_cold_**)(**_warm_**)] air advection was occurring behind the warm front. Nashville, TN and Charleston, SC also had this pattern of air advection.

28. As the frontal system from western Kentucky to the Texas Gulf Coast advances generally southeastward, stations ahead of the cold front at map time can expect temperatures to [(**_rise_**)(**_fall_**)] after the front passes their location. The advance of the colder air is suggested by the sequence of temperatures shown from Dallas (36 °F) to Oklahoma City (30 °F) to Wichita (25 °F) to North Platte (7 °F).

As the calendar approached the spring season, successive pushes of colder air became less frequent, less frigid, and less likely to reach lower latitudes so that temperatures rose as systems provided more frequent warm-air advection. Also, increasing daily doses of sunshine allowed the ground to warm. The combination of more solar radiation and more frequent and stronger warm air advection marked the change of seasons.

Suggestions for further activities: As weather systems cross your region, you might call up the "Isotherms, Fronts, & Data" map on the course website and identify patterns of cold or warm air advection associated with these passing systems. Warm and cold air advection patterns are often best seen in spring and autumn seasons when clashes of air masses make for dramatic weather episodes. You can lightly color regions of warm (red) and cold (blue) air advection on the map and relate them to the weather systems and to your daily temperature patterns.

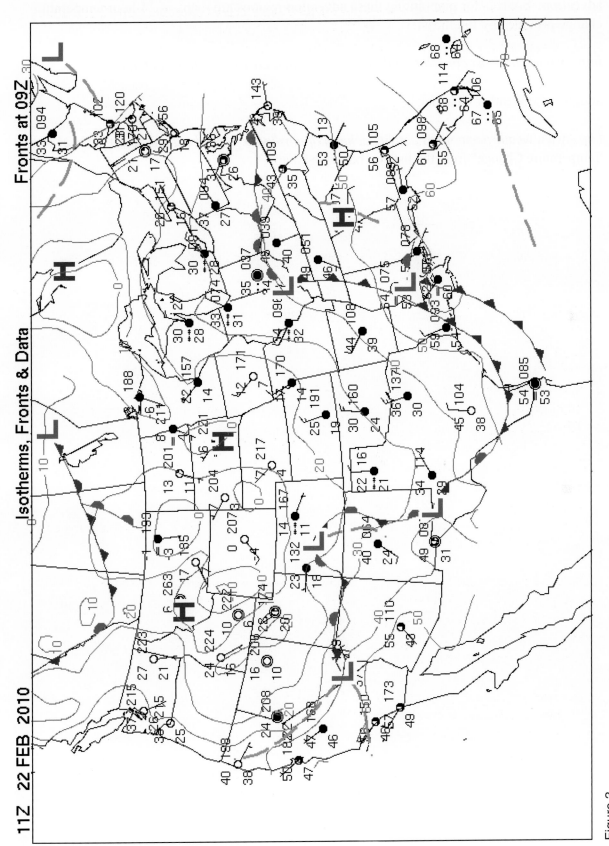

Figure 3.
Map of Isotherms, Fronts & Data for 12Z 22 FEB 2010.

Large temperature changes from one day to the next are likely to be the result of air mass advection. Sources for pinpointing these advection regions are maps of 24-hour temperature change provided by Intellicast:

http://www.intellicast.com/National/Temperature/delta.aspx?location=default

and, by The Weather Channel:

http://www.weather.com/maps/activity/achesandpains/index_large.html and select "US: 24 hour Temperature Change" from the list box below the map.

Investigation 4B:

HEATING AND COOLING DEGREE-DAYS AND WIND CHILL

Objectives:

Weather, by definition, refers to the state of the atmosphere mainly in terms of its effect upon life and human activities. Perhaps the most noticeable aspect of weather to individuals is air temperature. Outside air temperatures can be such that energy usage is necessary to make interior living spaces comfortable. The chilling effect of temperature combined with wind on exposed flesh is another aspect of weather that people in colder climates must guard against.

After completing this investigation, you should be able to:

- Calculate the number of heating or cooling degree-days accumulated on a given day, and demonstrate the use of current data to determine the number of heating or cooling degree-days in selected locations.
- Describe the pattern of average annual heating-degree totals over the coterminous United States.
- Determine the wind chill temperature based on temperature and wind observations.

Introduction:

1. During cool or cold weather episodes, fuel is consumed to make buildings comfortable living spaces. A useful indicator of fuel consumption for heating purposes is the determination of heating degree-days. They are calculated by accumulating one heating degree-day (HDD) unit for each Fahrenheit degree the daily **mean** (average) **temperature** is <u>below</u> the base value of 65 °F (18 °C). For example, a day with a maximum temperature of 70 °F and a minimum of 50 °F has a mean temperature of [...] subtracting 60 from 65 yields 5 heating degree-day units for that day. Hence, [...] h a high temperature of 40 °F and a low of 20 °F produces **[(_20_)(_35_)(_40_)(_65_)]** [...] egree-days (HDD).

[...] anying **Figure 1** map displays the <u>average annual total</u> number of heating [...] accumulated at various locations around the country. Assume that each location [...] er of the number plotted on the map. **Determine the pattern of degree-days accumulated yearly by drawing on the map contour lines representing 2000, 4000, 6000, 8000, and 10,000 heating degree-days.** Be sure to label each contour.

2. Compare your map analysis to the map appearing in **Figure 2**, which is based on data from many additional locations. According to Figure 2 map, Southern **[(_California_)(_Texas_) (_Florida_)]** has the lowest annual heating degree-day totals among the lower-48 states.

3. Latitude, elevation, and nearness to large bodies of water are factors that influence the annual heating degree-day pattern appearing on the map you analyzed. Of these three

factors, [(*latitude*)(*elevation*)(*nearness to large water bodies*)] is generally the most important for the vast majority of the coterminous U.S.

4. Examine **Figure 3**. Figure 3 shows the annual temperature curves for (A) San Francisco, CA, a maritime station and (B) St. Louis, MO, a continental location. They both have nearly the same annual mean temperature as represented by the dashed line. Draw a straight line across the graph representing the 65 F° isotherm. Although both locations are at about the same latitude, the city that accumulates heating degree-days <u>during more months</u> of the year is [(*St. Louis*)(*San Francisco*)].

5. The city that records more heating degree-days <u>annually</u> is [(*San Francisco*)(*St. Louis*)]. (We suggest you locate San Francisco and St. Louis on the Figure 2 map to confirm your response. This result is typical for West coast maritime locations compared to continental locations at the same latitude. Even though heating degree days may be accumulated at continental locations over a shorter part of the year, much lower temperatures during the cold part of the year cause rapid accumulation of heating degree-days.)

6. When the weather is too hot for human comfort, there is a need to cool air in living spaces. Cooling degree-days are calculated to estimate fuel needs for air conditioning when the day's mean temperature rises <u>above</u> 65 °F. The calculation is made by subtracting 65 °F from the day's mean temperature. On a day when the maximum temperature is 95 °F and the minimum temperature is 75 °F, the number of cooling degree-days (CDD) produced is [(*20*)(*65*)(*75*)(*95*)].

7. It takes only a short time outdoors in cool or cold weather for a person to realize that temperature and wind both play major roles in the rate at which the human body loses heat to the environment. The National Weather Service Windchill Chart (**Figure 4**) enables us to determine the windchill temperature index (WCT) based on combined effects of air temperature and wind speed. The WCT is an attempt to approximate the rate of sensible heat loss from exposed skin caused by the combined effect of low air temperature and wind. However, when the air is calm and there is no additional heat-loss effect from wind, one would expect the WCT to be [(*higher than*)(*the same as*) (*lower than*)] the existing air temperature. Actually, windchill temperatures are calculated only when winds are above 3 miles per hour.

8. According to the Windchill Chart, the WCT is [(*–5*)(*–2*)(*–17*)(*–19*)] °F when the temperature is 5 °F and the wind speed is 30 miles per hour (mph).

9. It can be seen from the Windchill Chart that above 35 mph, changes in wind speed have relatively little effect on WCT values. At 25 °F, an increase in wind speed from 5 to 10 mph causes a 4-degree drop in the WCT value. At the same temperature, an increase in wind speed from 40 to 45 mph reduces the WCT by [(*6*)(*5*)(*3*)(*1*)] Fahrenheit degree(s).

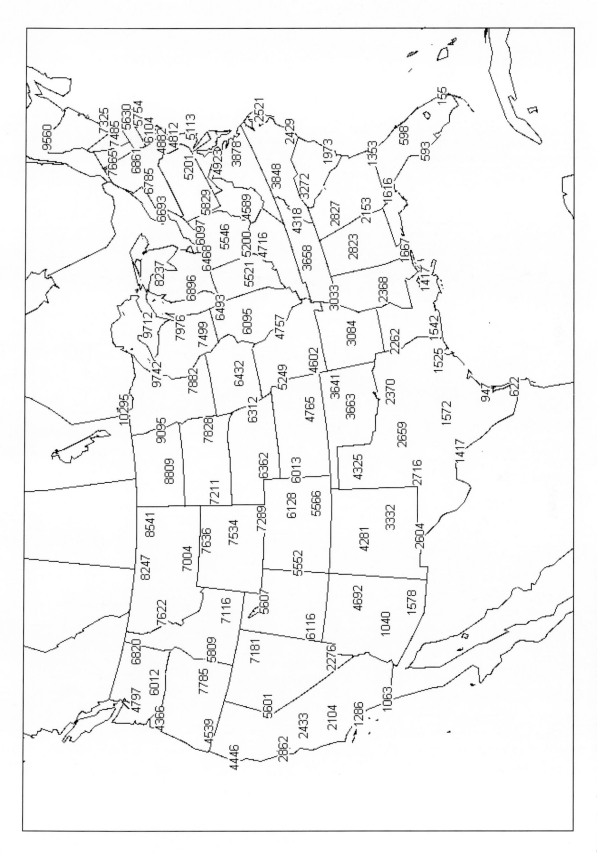

Figure 1. Average total annual heating-degree days.

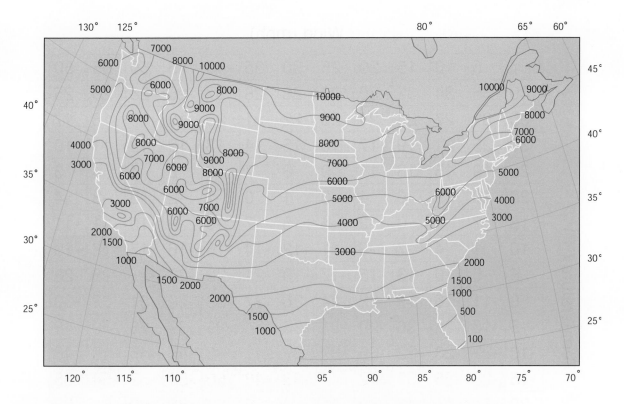

Figure 2.
Average annual heating-degree day total over the lower 48 states using a base of 65 °F.

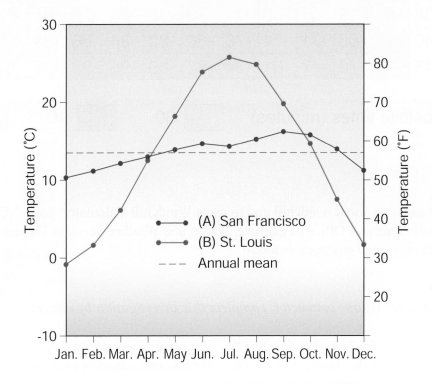

Figure 3.
Variation of average monthly temperature for (A) San Francisco and (B) St. Louis.

Wind (mph)

Temperature (°F) \ Calm	5	10	15	20	25	30	35	40	45	50	55	60
40	36	34	32	30	29	28	28	27	26	26	25	25
35	31	27	25	24	23	22	21	20	19	19	18	17
30	25	21	19	17	16	15	14	13	12	12	11	10
25	19	15	13	11	9	8	7	6	5	4	4	3
20	13	9	6	4	3	1	0	-1	-2	-3	-3	-4
15	7	3	0	-2	-4	-5	-7	-8	-9	-10	-11	-11
10	1	-4	-7	-9	-11	-12	-14	-15	-16	-17	-18	-19
5	-5	-10	-13	-15	-17	-19	-21	-22	-23	-24	-25	-26
0	-11	-16	-19	-22	-24	-26	-27	-29	-30	-31	-32	-33
-5	-16	-22	-26	-29	-31	-33	-34	-36	-37	-38	-39	-40
-10	-22	-28	-32	-35	-37	-39	-41	-43	-44	-45	-46	-48
-15	-28	-35	-39	-42	-44	-46	-48	-50	-51	-52	-54	-55
-20	-34	-41	-45	-48	-51	-53	-55	-57	-58	-60	-61	-62
-25	-40	-47	-51	-55	-58	-60	-62	-64	-65	-67	-68	-69
-30	-46	-53	-58	-61	-64	-67	-69	-71	-72	-74	-75	-76
-35	-52	-59	-64	-68	-71	-73	-76	-78	-79	-81	-82	-84
-40	-57	-66	-71	-74	-78	-80	-82	-84	-86	-88	-89	-91
-45	-63	-72	-77	-81	-84	-87	-89	-91	-93	-95	-97	-98

Frostbite times (minutes)   30   10   5

Figure 4.
NWS Windchill Chart.

For more information about windchill and to use a Windchill Calculator, go to NOAA's National Weather Service Office of Climate, Water, and Weather Services Windchill website at: *http://www.nws.noaa.gov/om/windchill/*.

As directed by your course instructor, complete this investigation by either:

1. *Going to the Current Weather Studies link on the course website, or*
2. *Continuing to the Applications section for this investigation that immediately follows in this Investigations Manual.*

Investigation 4B: Applications

HEATING DEGREE-DAYS AND WIND CHILL

Heating and Cooling Degree Days and wind chill values are calculated with the use of observational data collected at weather stations. Annual totals of Heating Degree Days are accumulated beginning 1 July while Cooling Degree Days begin with 1 January.

10. **Figure 5** is a NWS map of maximum and minimum temperatures, in degrees Fahrenheit, at selected stations across the U.S. during the 24-hour period ending at 12Z on Tuesday, 23 February 2010. The red upper number is the period's maximum temperature and the blue lower number the minimum temperature at the station denoted by the green dot. For the contiguous U.S., not surprisingly, temperatures in the high plains of the north-central U.S. were among the coolest with Fargo, on the ND-MN border, being the station on this map reporting the *lowest* <u>maximum</u> temperature for the date. The maximum temperature reported at Fargo during the 24 hours was **[(*24*)(*20*)(*17*)]** °F. The minimum temperature there was –1 °F. As noted in the **Applications** section of Investigation 4A, a cold air mass had been invading that region following the cold front bringing cold air advection to stations in the region.

11. The *lowest* <u>minimum</u> temperature on the map of the contiguous U.S. was –2 °F reported at **[(*Minneapolis, MN*)(*North Platte, NE*)(*Trinidad, CO*)]**.

12. On the warmer side, the *highest* <u>maximum</u> temperature was a balmy 79 °F at Ft. Myers, Florida while their minimum was 64 °F. However, the *highest* reported <u>minimum</u> temperature for the period in the contiguous U.S. occurred at Key West, Florida, where it only dropped down to **[(*75*)(*72*)(*68*)]** °F.

13. Meanwhile, the desert Southwest was not quite as warm as Florida, as typified by Phoenix, Arizona, which had a maximum and minimum of 61 and 41 °F, respectively. Based on Phoenix's maximum and minimum, the mean temperature at Phoenix was **[(*58*) (*51*)(*49*)]** °F.

14. The mean temperature derived from the maximum and minimum temperatures at Phoenix indicates **[(*heating*)(*cooling*)]** degree day units were accumulated.

15. The number of degree days accumulated in Phoenix was **[(*4*)(*14*)(*25*)]**.

Refer again to Figure 5 to answer the following questions, keeping in mind that the base of 65 °F is used to determine whether Heating Degree Day units or Cooling Degree Day units, if any, were produced and how many. Also, follow the NWS practice that, if necessary, the mean daily temperature is rounded *up* to the nearest whole degree before calculating HDD or CDD.

16. Fargo experienced a mean temperature of [(*14*)(*10*)(*6*)] °F.

17. Therefore, Fargo accumulated [(*35*)(*39*)(*42*)(*55*)] HDD units on that day.

18. Orlando, FL with a high of 75 °F experienced a mean temperature of [(*58*)(*62*)(*68*)] °F.

19. Orlando had 3 [(*HDD*)(*CDD*)] as a result.

20. Scanning the map's maximum and minimum temperatures, it can be inferred that utility energy usage in the contiguous U.S. during this daily period was probably more needed for *heating* across the [(*northeastern*)(*north-central*)(*southwestern*)(*southeastern*)] portion of the U.S. For real contrasts, you might note Alaska where Barrow to the north had -15 °F as both high and low while Anchorage to the south along the Gulf of Alaska was as warm as Dallas, TX!

21. The NWS map of highest and lowest temperatures can be accessed from the course website. Go to the website, scroll down to **Climate** and click on the "National Temps/ Precip." link. When viewing the NWS page with the latest map, go to the "Max/Min Temperatures" option bar to the upper left of the map and click on the down arrow. The list of temperature dates displayed shows that, including the current day's map, you can access max/min temperature maps for a total of 7 days. Similarly, using the same option bar, one can highlight any of the current or past six days' maps of [(*precipitation*)(*wind speed*)] which can be acquired by then pressing GO. The option bar to the upper right provides future climate outlooks.

Also on the course website, **Climate** section, there is a link to "Local NWS Offices". That link provides a map of local National Weather Service Offices. Click on the yellow circle of your closest NWS Office; the webpage of that NWS Forecast Office appears. In the menu along the left side of the page is an option for "Climate" information and a link to "Local" for climatic information in various forms for one or more cities in the office's service area. Choosing (1) Product: *Daily Climate Report (CLI)*, (2) Location, (3) Most Recent (yesterday's), or archived choice, then (4) GO. The new page will provide the requested day's climatic data for the station showing maximum, minimum and average temperatures as well as heating and cooling degree days for that date and the season. You can see your local data (without having to do the arithmetic!) for the season.

HDD values and their effect on homeowners' heating bills are an issue for many as heating costs are yet another stress to tight budgets.

22. Heating and Cooling Degree Day units reflect energy needs to attain indoor human comfort. Outdoors the cooling effect of wind along with temperature on an individual is reflected by using the windchill equivalent temperature. The Figure 5 map shows that the minimum temperature in central North Dakota was about 0 °F. Assume that central ND was experiencing a wind speed of 10 miles per hour when the temperature was 0 °F. Using the windchill equivalent temperature (WET) Table in Figure 4 of this Investigation

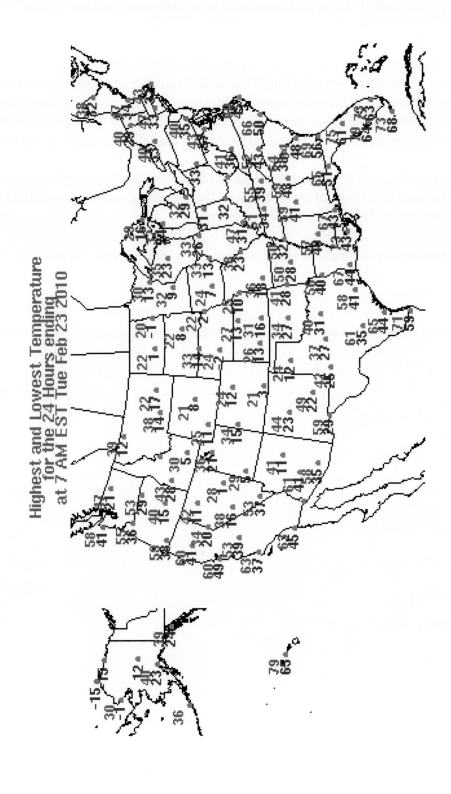

Highest and Lowest Temperature
for the 24 Hours ending
at 7 AM EST Tue Feb 23 2010

Figure 5.
Map of maximum and minimum temperatures at selected stations for the 24-hour period ending at 12Z on Tuesday, 23 February 2010.

4B, the WET for this combination of temperature and wind speed would have been [(_–4_)(_–11_)(_–16_)(_–22_)] °F. According to the Table's shading, an increase of wind speed to 15 mph with the same 0 °F temperature would bring danger of frostbite within 30 minutes!

Suggestion for further activities: High energy prices, along with increasing evidence of climate change due to human activity, highlight the need for greater awareness concerning energy use. Space heating and air conditioning are major consumers of energy (and financial resources). Tracking the accumulation of HDDs and CDDs where you live can lead to more informed decision-making and actions related to energy consumption for heating and cooling.

The HDD values you determined in this investigation are reasonably consistent in pattern with the national annual HDD map of the first part of the investigation. Tables of HDD for selected cities to the end of each month can be found at:

http://www.cpc.ncep.noaa.gov/products/analysis_monitoring/cdus/degree_days/mctyhddy.txt

and tables of CDD at:

http://www.cpc.ncep.noaa.gov/products/analysis_monitoring/cdus/degree_days/mctycddy.txt

Investigation 5A:

AIR PRESSURE CHANGE

Objectives:

Atmospheric pressure at any place on the Earth's surface changes over time. In midlatitudes, these changes are often related to the migration of air masses. An air mass is a huge volume of air covering thousands of square kilometers that is relatively uniform horizontally in temperature and water vapor concentration. Neighboring air masses with contrasting characteristics tend not to mix so that distinct boundaries (fronts) form between them. Fronts occupy troughs of low pressure. As air masses travel over Earth's surface, there are changes in the air pressure and weather at places in their paths. Those changes are particularly dramatic at or near the moving air mass boundaries.

After completing this investigation, you should be able to:

- Identify air pressure changes and other local weather conditions that indicate the passage of a cold front.
- Relate local air pressure changes and weather conditions to the presence of different air masses before and after the passage of a cold front.
- Estimate the speed of movement of a strong, well-defined cold front.

Introduction:

Highs (**H**s) and Lows (**L**s) plotted on surface weather maps identify the highest or lowest pressure centers of broad-scale pressure systems. Up to several of these might be seen on a map of the coterminous United States at any particular time. These systems generally move from west to east across our midlatitude portion of the globe. Pressure values at locations in the paths of these migrating systems fall as a Low approaches or a High departs, and pressures rise with approaching Highs or departing Lows. Fronts can mark the boundaries of these Highs and generally show the transition from one air mass to another. Fronts frequently anchor at low-pressure centers.

Figure 1 is a cross-section schematic of a cold front moving from west to east in the drawing. The vertical scale of the cross-section is greatly exaggerated for clarity. The warm (red) air mass is being replaced by the cold (blue) air mass. Each air mass would have a high-pressure center as shown. Imagine being located where the air pressure is highest in the warm air mass. You would find that pressure, temperature, and other weather parameters vary as the front approaches and moves past your location. (For a map view, see Figure 2 of Investigation 4A.)

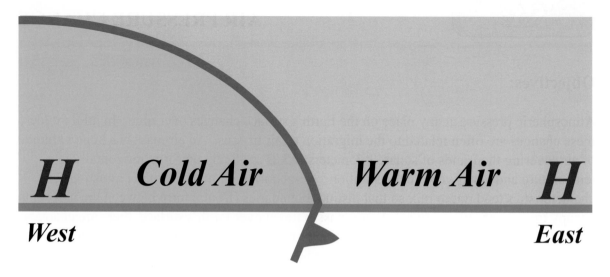

Figure 1.
Cross-section of idealized cold front.

1. As the cold front passes your location, the temperature where you are would [(*rise*) (*remain steady*)(*fall*)].

2. As the cold front moves toward and passes your location, the air pressure would [(*rise then fall*)(*fall then rise*)].

As directed by your course instructor, complete this investigation by either:

1. *Going to the Current Weather Studies link on the course website, or*
2. *Continuing to the Applications section for this investigation that immediately follows in this Investigations Manual.*

Investigation 5A: Applications

AIR PRESSURE CHANGE

The 2009-2010 winter season was unusual in many parts of the U.S. Storm systems delivered record snowfall amounts to many East Coast cities. The central U.S. had its share of snowy conditions while rains were extensive on the West Coast. These results were consistent with the existence of the El Niño pattern in the tropical Pacific Ocean at the time and its effects on U.S. weather. We will consider the El Niño phenomenon in a later investigation. One consequence of these weather patterns was the storm track, the persistent path of low-pressure systems, across the Gulf Coast region. Over the last weekend in February one such Low traveled across the area.

Figure 2 is the surface weather map for 12Z 27 FEB 2010 (7 AM EST Saturday morning). At map time, the storm system was marked by a low-pressure center in the eastern Gulf of Mexico as depicted by the **L** south of Alabama and the Florida peninsula. From the low-pressure center a warm front curved to the east and southeast while a cold front arced southward and southwestward.

3. Examine the station model for Miami, Florida, located on southeast Atlantic coast. The weather conditions at Miami at map time were: temperature 52 °F, dewpoint 43 °F, partly cloudy skies, coded pressure value "132" and calm wind. At map time the southeast end of the warm front was shown just touching the west coast of southern Florida. From the frontal symbols and the map locations shown, at 12Z the warm front [(***had already***) (***had not yet***)] passed Miami's location.

4. Note the weather conditions plotted at Key West, FL, at the southern end of the Florida Keys islands. Compare the temperatures at Key West and those of Miami. The air ahead of the front was [(***cooler***)(***warmer***)].

5. Also compare the dewpoints at those same stations ahead of and behind the front. The air ahead of the front contained [(***more***)(***less***)] water vapor as reflected by the dewpoint values.

Figure 3 is the meteogram for Miami, FL (MIA) for the 24-hour period from 1000Z 27 FEB 2010 (*100227/1000* in the top heading of the meteogram) to 1000Z 28 FEB 2010. [Meteograms for current weather conditions for the preceding 24-hour period at selected stations can be obtained from the course website (**Surface** section, *Meteogram for Selected Cities*).] On the Figure 3 meteogram, draw a vertical line on the meteogram at the 12Z 27 FEB 2010 time (labeled "27/12" tick mark) corresponding to the Figure 2 surface weather map time (Z times along the lower graph axis). Label this time as "Fig. 2".

6. The weather conditions at Miami on the Figure 2 surface weather map [(***were***)(***were not***)] the same as those depicted on the Figure 3 meteogram at that time.

7. From 12Z to 13Z on 27 FEB, the air pressure at Miami [(*__fell__*)(*__rose__*)]. Following 13Z the pressure rose slightly and then was steady from 14Z to 15Z.

8. At 13Z the wind direction at Miami was from the east-southeast. From 14Z to 16Z the winds were generally from the [(*__south-southeast__*)(*__southwest__*)(*__northwest__*)]. While this was a small change, it was of moderate intensity and consistent in direction.

9. From 12Z to 20Z on 27 FEB, the temperature generally [(*__fell__*)(*__rose__*)], first rapidly for three hours, then held steady from 16Z to 19Z before rising more gradually. The initial three hours (12 to 15Z) occurred from 7 to 10 a.m. local time when normal daily heating occurs, but the wind direction would also suggest warm air advection might be occurring as well.

10. From 12Z to 16Z, the dewpoint [(*__fell__*)(*__rose__*)].

11. Between 12Z and 16Z, the lowering pressure, relatively rapidly rising temperature and rising dewpoint with wind directions changing from a southeasterly to more southerly direction suggest that the [(*__cold__*)(*__warm__*)] front may have passed Miami between 13Z and 14Z. This would be consistent with the front's analyzed position on the 12Z surface map.

12. Next examine the period from 21Z to 22Z on the meteogram. Prior to 21Z air pressure was [(*__rising__*)(*__falling__*)]. Following 21Z the pressure rose relatively quickly for an hour, then steadied and finally rose more gradually from 23Z until the end of the meteogram period.

13. Between 21Z and 22Z, the wind shifted direction from being southwesterly to being from the [(*__south-southeast__*)(*__southwest__*)(*__northwest__*)]. Wind directions remained generally from that direction later in the meteogram period.

14. The temperature [(*__rose slowly__*)(*__remained steady__*)(*__fell rapidly__*)] from 21Z to 22Z. The dewpoint was steady during this hour but then decreased in the following hour and eventually both the temperature and dewpoint were at values near the end of the meteogram period similar to the beginning. Also, note that at 22Z, Miami recorded a rain shower during overcast conditions.

15. The pattern of changes in temperature, dewpoint, wind direction and air pressure between 21Z and 22Z were consistent with the passage of a [(*__warm__*)(*__cold__*)] front.

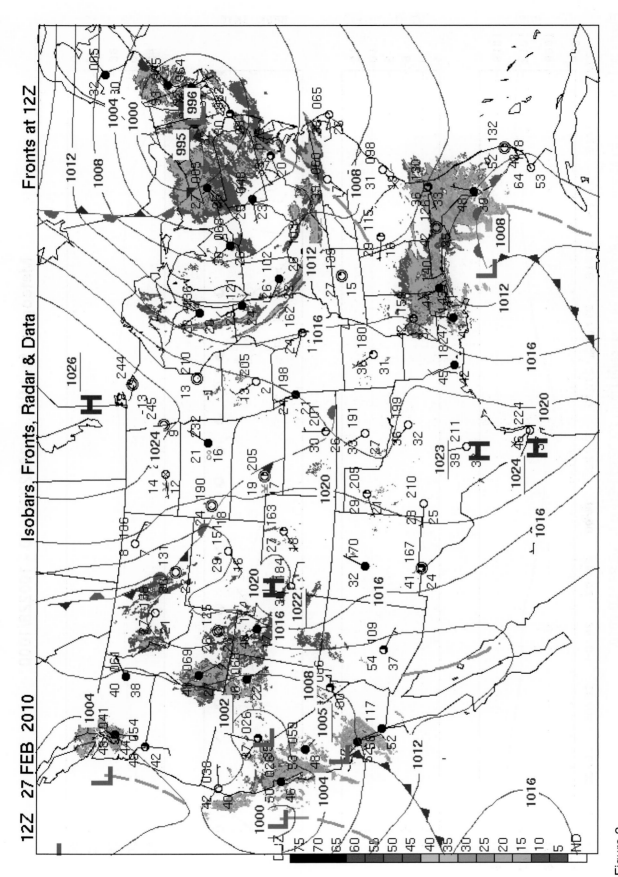

Figure 2.
Surface weather map for 12Z 27 FEB 2010.

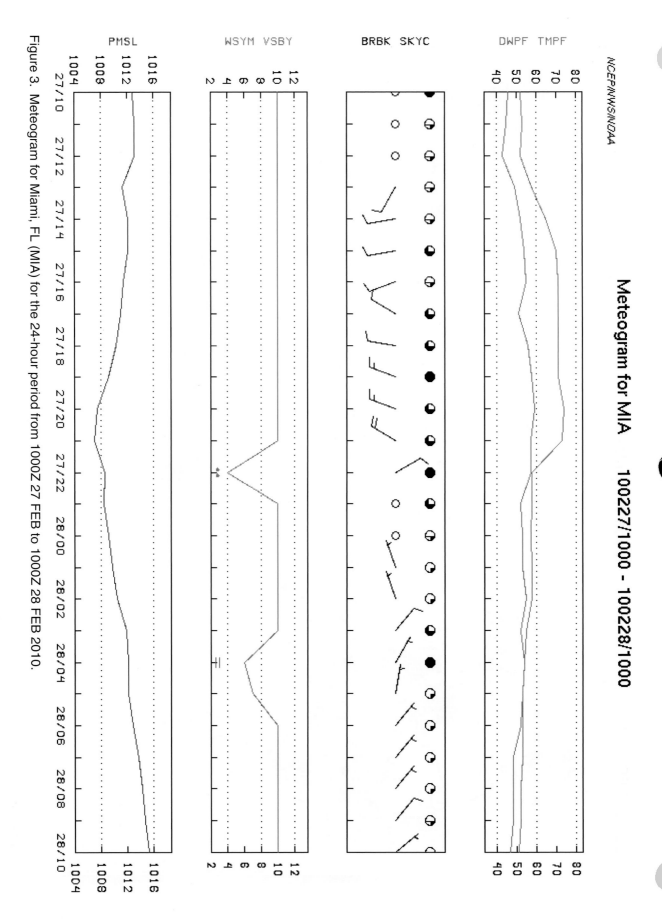

Figure 3. Meteogram for Miami, FL (MIA) for the 24-hour period from 1000Z 27 FEB to 1000Z 28 FEB 2010.

Figure 4 is the surface weather map for 00Z 28 FEB 2010, twelve hours following the Figure 2 map. On the Figure 3 meteogram, draw a vertical line on the meteogram at the 00Z 28 FEB 2010 time (tick mark at 28/00) corresponding to the Image 3 surface weather map time. Label this time as "Fig. 4".

16. Note the temperature, dewpoint, sky cover, wind direction and pressure at Miami at 00Z 28 FEB. These values [(**_were_**)(**_were not_**)] the same as those shown on the 00Z, Figure 4 surface map.

17. From the cold front's position shown on the Figure 4, 00Z 28 FEB map, the front at 00Z [(**_had already_**)(**_had not yet_**)] passed Miami's location.

18. This Figure 4 map depiction of frontal locations [(**_was_**)(**_was not_**)] consistent with the changes in weather conditions and the pressure pattern shown on the Figure 3 meteogram.

This period captured the change of weather conditions experienced at a location with a passing low-pressure system and its accompanying fronts on a single meteogram. These maps and the meteogram graphically display the principle that southern Florida really does have "weather" at least on the last day of (meteorological) winter!

You can acquire meteograms for the NWS station closest to you if not among those on the course website, (or to other locations around the U.S. and world) by going to: *http://vortex. plymouth.edu/statlog.html*. [This is the "click here" link on the bottom of the *Meteograms for Selected Cities in the U.S.* page from the course website **Surface** data section.]

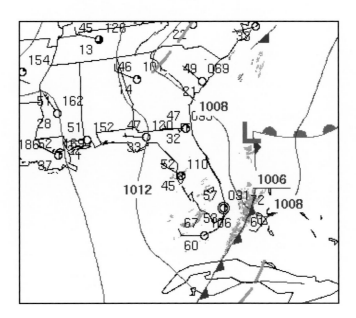

Figure 4.
Surface weather map for 00Z 28 FEB 2010.

ATMOSPHERIC PRESSURE IN THE VERTICAL

Objectives:

One of the most important properties of the atmosphere is air pressure. After barometer readings at Earth's surface are "reduced" or "corrected" to sea level, the resulting air pressure values still vary from place to place and with time. The principal reason for this unevenness in air pressure arises from the variability of air temperature over distance and time. With otherwise similar conditions, warm air is less dense than cold air. Density variations produced by temperature differences lead to pressure differences in the horizontal throughout the atmosphere.

This investigation simulates special "blocks" to study basic understandings about pressure and pressure differences produced by density variations.

After completing this investigation, you should be able to:

- Explain what air pressure is.
- Explain how variations in air temperature cause differences in air pressure.
- Describe how density contrasts between warm and cold air produce horizontal variations in air pressure at different altitudes in the atmosphere.

Introduction:

To study pressure, we must first define it. Pressure is a force acting on a unit area of surface (*e.g.*, pounds per square inch is a pressure measurement). Air pressure is described as the weight (a force) of an overlying column of air acting on a unit area of horizontal surface. To investigate the concept of pressure we will simulate the use of tall and short "blocks". One tall block and one short block are shown in **Figure 1**. Tall blocks are cube-shaped and short blocks have the same size base as the tall blocks and are half as high.

Whether tall or short, the blocks employed in this investigation have the following common characteristics:

a. All blocks have the same weight regardless of the volume they occupy.
b. All blocks have the same size square base.
c. All individual blocks exert the same downward pressure on the surface beneath them (because the equivalent weights are acting on the same size bases).

1. **Figure 1** shows one tall red block and one short blue block side-by-side on their square bases on the flat horizontal surface of a table (**T**). Because both blocks weigh the same (although they have different volumes) and their bases are the same size, the blocks exert [(*__equal__*)(*__unequal__*)] pressure on the surface of the table.

2. The shorter blocks occupy half of the volume of the taller blocks while containing equal masses. (We know this because they weigh the same.) Because density is a measurement of mass per unit volume, the smaller blocks are [(*__twice__*)(*__half__*)] as dense as the larger blocks.

3. In **Figure 2**, another identical block was placed on top of each block already on the table. Each stack is now exerting [(*__the same__*)(*__twice the__*)] amount of pressure on the table as the single blocks did initially.

4. The pressure exerted on the table by the tall stack is [(*__equal__*)(*__not equal__*)] to the pressure exerted on the table by the short stack.

5. As shown in **Figure 3**, the two stacks are side-by-side with another identical block added to each stack (for a total of 3 blocks in each stack). An imaginary surface (**1**) has been inserted horizontally through the two stacks so that two shorter blocks and one taller block are positioned beneath the surface. Compare the pressure exerted on the imaginary surface by the overlying blocks. The taller-block stack exerts [(*__greater__*)(*__equal__*)(*__less__*)] pressure on this imaginary surface than does the shorter-block stack.

6. **Figure 4** shows two more blocks added for a total of five in each stack. A second imaginary horizontal surface (**2**) is added beneath the top short block and the three top tall blocks. The pressure exerted on the table (T) by the tall stack is [(*__equal__*)(*__unequal__*)] to the pressure exerted on the table by the short stack.

7. On the lower imaginary surface (**1**) in Figure 4, the pressure exerted by all the overlying short blocks is [(*__one-half__*)(*__three-fourths__*)(*__the same as__*)] the pressure exerted by all the overlying tall blocks.

8. On the top imaginary surface (**2**) in Figure 4, the downward pressure exerted by the overlying short block is [(*__equal to__*)(*__one-half__*)(*__one-third__*)] the pressure exerted by the overlying tall blocks.

9. Starting at the table top and moving upward in Figure 4, the <u>difference</u> in downward pressure on imaginary horizontal surfaces exerted by the overlying portions of the two stacks [(*__increases__*)(*__decreases__*)].

10. In the [(*__taller, less dense__*)(*__shorter, more dense__*)] stack, the pressure decreases more rapidly with height.

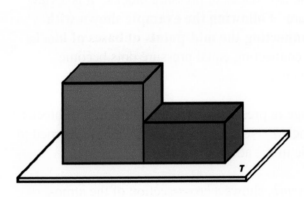

Figure 1.
One tall, one short block.

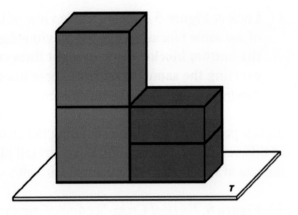

Figure 2.
Two tall, two short blocks.

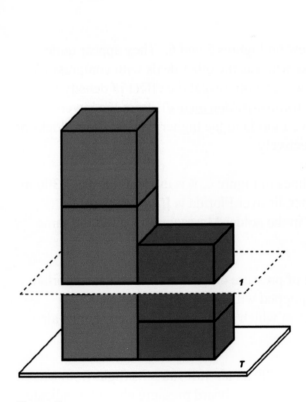

Figure 3.
Three tall, three short blocks with surface *1* inserted.

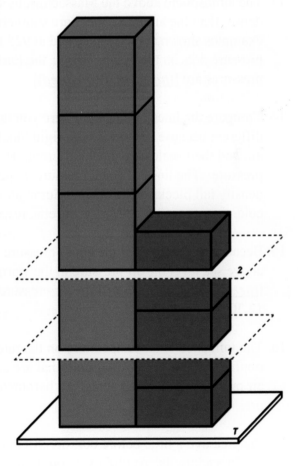

Figure 4.
Five tall and short blocks with surfaces *1* and *2* inserted.

11. Look at **Figure 5** showing a side view of the two stacks of pressure blocks. It is a view of the same blocks seen in the previous figure. **Following the example shown with the bottom blocks, draw straight lines connecting the mid-points of bases of blocks exerting the same pressures.** These lines connecting equal pressure dots become [(*more*)(*less*)] inclined with an increase in height.

To this point we have been examining the change in pressure with height in stacks of blocks of different density (short blocks versus tall blocks). Now we apply what we have learned to the rate at which air pressure drops with altitude in the atmosphere.

12. **Figure 6**, *Vertical Cross-Section of Air Pressure*, shows a cross-section of the atmosphere based on upper-air soundings obtained by radiosondes simultaneously at Miami, Florida and at Chatham, Massachusetts, approximately 1250 mi. (2000 km) apart on 00Z 28 January 2009. Air pressure values in millibars (mb) are plotted as marks at the altitudes where they were observed, starting with identical values (1000 mb) at the Earth's surface. Over Miami, the atmosphere was exerting a pressure of 200 mb at an altitude of approximately [(*11,700*)(*12,000*)(*12,300*)] m above sea level.

13. The atmosphere above the Massachusetts weather station was colder and therefore denser than the air above the more southern and warmer Florida location. Following the examples shown at the surface and at 925 mb, draw straight lines connecting equal air-pressure dots on the graph. Above the Earth's surface these lines representing equal air pressures are [(*horizontal*)(*inclined*)].

14. Compare the lines of equal pressure you drew on Figures 5 and 6. They appear quite different because one deals with rigid blocks whereas the other deals with compressible air, and their scales are much different. However, both reveal the effect of density on pressure. The lines of equal pressure slope [(*upward*)(*downward*)] from the lower-density tall blocks or warm air column above Florida to the higher-density short blocks or cold air column above Massachusetts, respectively.

15. Because of the slope of the equal-pressure lines in Figure 6, it is evident that at 12,300 m above sea level the air pressure in the warmer air over Florida is [(*higher than*) (*the same as*)(*lower than*)] the air pressure in the colder Massachusetts air at the same 12,300-m altitude.

16. The influence of air temperature on the rate of pressure drop with altitude has important implications for pilots of aircraft that are equipped with air pressure altimeters. An air pressure altimeter is actually a barometer in which altitude is calibrated against air pressure.

Imagine that a little before 00Z on 28 January 2009 an aircraft starts its flight from Miami to Massachusetts. At 00Z over southern Florida, the onboard pressure altimeter indicates that the aircraft is at 5800 meters above sea level. From Figure 6, the air pressure is about [(*500*)(*400*)] mb at an altitude of 5800 meters over Florida.

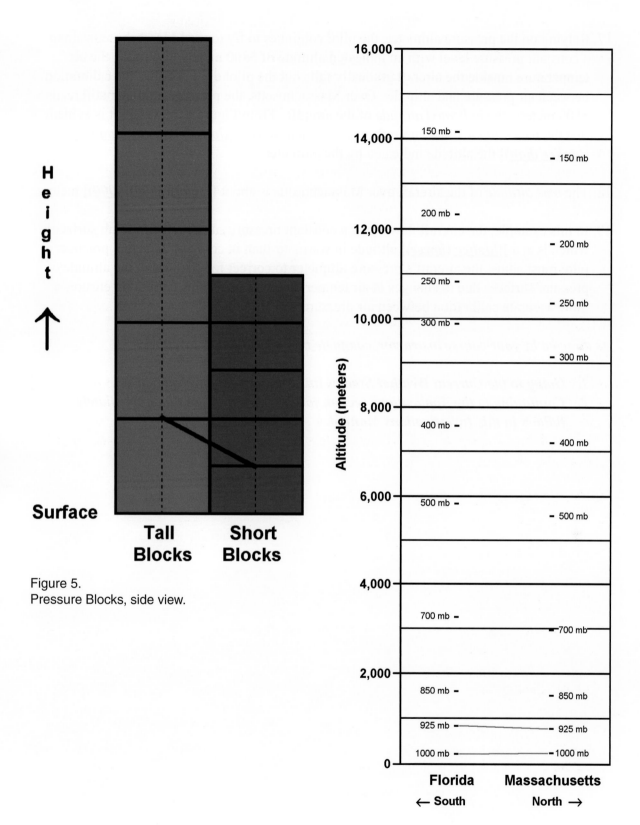

Figure 5.
Pressure Blocks, side view.

Figure 6.
Vertical Cross-Section of Air Pressure

17. Relying on the pressure altimeter, the pilot continues to fly toward Massachusetts along a constant pressure level with an <u>indicated</u> altitude of 5800 meters. En route, the air temperature outside the aircraft gradually falls but the pilot does not alter the calibration between air pressure and altitude. Over Massachusetts, the pressure altimeter still reads 5800 meters, the *indicated altitude* of the aircraft. From Figure 6, however, it is evident that the *true altitude* of the aircraft over Massachusetts is **[(*lower than*)(*the same as*) (*higher than*)]** the altitude indicated by the altimeter.

18. The *true altitude* of the aircraft over Massachusetts is about **[(*5600*)(*5800*)(*6000*)]** meters.

19. In this example, the aircraft flew along a constant pressure surface (the 500-mb surface) which is at a **[(*higher*)(*lower*)]** altitude in warm air than in cold air. In actual practice, a pilot must adjust the aircraft's pressure altimeter to correct for changes in the altitude of pressure surfaces due to changes in air temperature en route. This correction ensures a more accurate calibration between air pressure and altitude.

As directed by your course instructor, complete this investigation by either:

  1. *Going to the Current Weather Studies link on the course website, or*
  2. *Continuing to the Applications section for this investigation that immediately follows in this Investigations Manual.*

Investigation 5B: Applications

ATMOSPHERIC PRESSURE IN THE VERTICAL

Figure 7 is the surface map for 00Z 3 MAR 2010. At that time a strong storm was moving up the East Coast and was centered over Cape Hatteras, North Carolina. With its northeastward movement and its internal counterclockwise circulation bringing air from off the Atlantic ahead of it, from the northeast, it is termed a *nor'easter*. Following that storm system the midsection of the U.S. was under the influence of high pressure. This extension of high pressure from central Canada to the western Gulf of Mexico is called a *ridge* of high pressure. Although this high-pressure pattern provided relatively fair weather, the temperatures varied from a balmy 60 °F at Brownsville in extreme southern Texas to a cool 31 °F at Green Bay, Wisconsin along Lake Michigan.

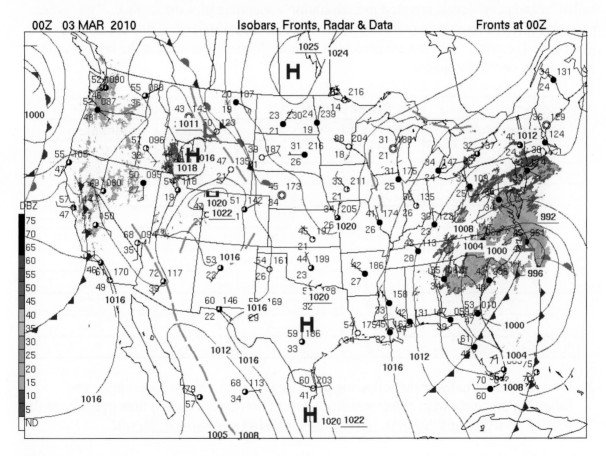

Figure 7.
Surface weather map for 00Z 3 March 2010. [NCEP/NOAA]

Radiosonde data can be employed in a variety of ways to provide meteorologists with powerful analytical tools. Data at various atmospheric levels above a station can be plotted in the radiosonde profile on the Stüve diagram that we examined in Investigation 2B. Here we consider atmospheric temperature and pressure conditions above the two stations mentioned above, which are about 2200 km apart. Radiosonde observations at 0000Z 3 March 2010 (*100303/0000*) from Brownsville (BRO), TX, and Green Bay (GRB), WI, depicting the tropospheric conditions over those stations are plotted on Stüve diagrams shown in **Figure 8** and **Figure 9**, respectively.

20. Trace with heavy pencil lines or a highlighter the temperature profiles (plotted curves to the right on the Stüve diagrams) for each station. Comparing the two temperature curves, the atmosphere (up to about 225 mb) was cooler above **[(*Green Bay*)(*Brownsville*)]**. (Note: This can be seen by laying one Stüve printout over the other and holding up to the light.)

The following table lists a portion of the text data from the two radiosonde observations. [Complete data for the most current soundings are routinely available from the course website section, **Upper Air**, "Upper Air Data - Text".] Pressure levels given are the so-called "mandatory" levels reported in each station's sounding plus the surface. The surface pressures differed somewhat (partially due to their altitude differences) and are listed as the first (lowest) entries. The data are presented with the highest pressures at the bottom as is found in the open atmosphere.

Pressure (mb)	Brownsville (BRO)		Green Bay (GRB)	
	Temp (°C)	Altitude (m)	Temp (°C)	Altitude (m)
100	−72.5	16380	−54.5	16000
200	−57.5	12210	−50.7	11520
300	−34.5	9520	−52.7	8900
400	−18.5	7440	−39.9	6990
500	−8.9	5740	−28.9	5430
700	0.8	3087	−11.3	2935
850	3.0	1515	−7.7	1436
925	9.6	823	−3.1	772
993 (sfc)	---	---	2.0	214
1019 (sfc)	19.2	7	---	---

21. Compare the altitudes of the following pressure levels on the two soundings. The following pressures were at *lower* altitudes over Brownsville (BRO) than over Green Bay (GRB):   **[(*925 mb*)(*700 mb*)(*500 mb*)(*300 mb*)(*100 mb*)(*none of these*)]**.

22. Constant-pressure "surfaces" are those that can be imagined as surfaces in the atmosphere on which the air pressure is everywhere the same, for example, the 500-mb surface. Comparing the pressure surfaces from 925 mb to 100 mb, the top of the plotted soundings, indicates that those constant-pressure surfaces slope from the generally warmer air over Brownsville **[(*downward*)(*upward*)]** to the cooler air over Green Bay.

It may be noted that the coldest temperature of the soundings was found over

Brownsville. It is typical that lower latitude stations have colder temperatures at the boundary of the troposphere and stratosphere (tropopause) than higher latitude stations.

23. In general, as one moves toward Earth's poles, one would expect the altitudes of a particular constant-pressure surface to become [(*__higher__*)(*__lower__*)].

24. This altitude change of a given pressure surface as one moves poleward is in response to [(*__higher__*)(*__lower__*)] average temperatures in the underlying air columns.

25. Assume you were to fly from Green Bay to Brownsville maintaining a 30,000-foot (9144 m) altitude as indicated by your pressure altimeter. Using a pressure altimeter, you are actually flying along a constant-pressure surface. (To visualize this flight, you can compare the altitudes of the 300-mb levels of the two stations, a pressure that is found near 30,000 feet.) As you approach Brownsville, your aircraft would actually be at a [(*__higher__*) (*__lower__*)] altitude than that indicated by your altimeter when you were over Green Bay.

26. If you were to fly from Green Bay to Brownsville while maintaining an actual altitude of 30,000 feet, the air pressure outside your plane would [(*__gradually increase__*) (*__remain the same__*)(*__gradually decrease__*)].

27. The tropopause, the level where the temperature becomes nearly constant (isothermal) or begins to increase with height (inversion), marks the top of the troposphere and can be seen in both Stüve diagrams. The tropopause over Brownsville was reported at 115 mb while that over Green Bay was at 283 mb (provided from a different data source). The depth of the troposphere (surface to tropopause) is termed "thickness" and is related to the average temperature of the air column. The troposphere was "thicker" over [(*__Green Bay__*) (*__Brownsville__*)] where the average temperature in the troposphere was generally greater.

Changes in tropospheric conditions above a station with passing weather systems can also be tracked with Stüve diagrams. This is one example of the influence of a strong passing cold front on tropospheric temperatures. You might also compare quite different stations such as Hilo, Hawaii and Anchorage, Alaska, from the course website or a similar pair of warm/cold locations via the Plymouth State University page.

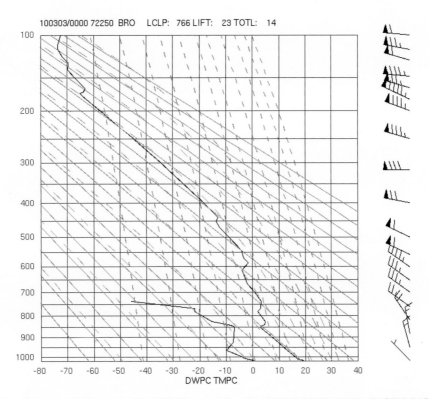

Figure 8. Stüve diagram of Brownsville (BRO) sounding for 00Z 3 MAR 2010. [NCEP/NOAA]

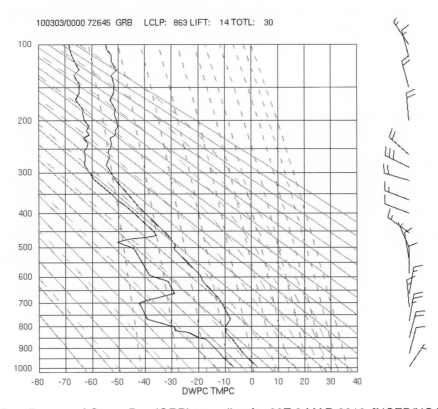

Figure 9. Stüve diagram of Green Bay (GRB) sounding for 00Z 3 MAR 2010. [NCEP/NOAA]

CLOUDS, TEMPERATURE, AND AIR PRESSURE

Objectives:

Clouds are an ever-present feature of Earth's atmosphere. A cloud is a visible suspension of tiny water droplets and/or ice crystals formed when water vapor condenses or deposits within the atmosphere. Air temperature changes resulting from air pressure changes play major roles in determining where clouds do and do not occur. Clouds develop where air ascends and dissipate where air descends.

After completing this investigation, you should be able to:

- Describe how air temperature changes as air pressure changes.
- Make clouds appear and disappear in a hypothetical bottle.
- Describe the role condensation nuclei play in enhancing cloud formation.
- Explain how most clouds form in the atmosphere.

Introduction:

Cloud formation and dissipation are closely related to temperature and pressure changes in the atmosphere. Vertical motions play a primary role as air rising or sinking in the atmosphere experiences pressure changes. These pressure changes, in turn, bring about temperature changes that can result in condensation and cloud formation, or evaporation and cloud dissipation.

The relationship between air pressure and temperature can be explored in the following thought demonstration. Place a thermometer, like a thin liquid crystal temperature strip, in a clean and dry empty (air-filled!) plastic 2-liter or larger beverage bottle sealed with its cap. A liquid crystal temperature strip works well because it is very sensitive to the temperature changes of its immediate environment, in this case, the surrounding air in the bottle. Secure the temperature strip with a piece of tape to hang at the center of the bottle. Screw the cap on tightly. [It is recommended that this experiment actually be done if possible. Temperature strips are available where aquarium supplies are sold, or at *http://www.ametsoc.org/amsedu/AERA/ed_mats.html*.]

After sealing the bottle and letting it rest for a minute or two, read the temperature. Exert pressure on the bottle so that its volume decreases. Reading the temperature strip after a few seconds have elapsed will show that the temperature of the air in the bottle rises. [A good way to squeeze the bottle to increase pressure on the air inside is to place the capped bottle so that about half of its length extends beyond the edge of a desk or table. Standing and with one hand on each end, push down on both ends of the sealed bottle so that it bends in the middle and partially collapses.]

Release the pressure so that the bottle expands again. It will be seen that as the bottle returns to its original shape, the air temperature in the bottle falls. [Be sure to hold the bottle in a squeezed position for at least a half-minute or so until the temperature stabilizes, then stop pushing down on the ends of the bottle. Try the bottle squeeze-and-release sequence several times while continuing to carefully observe temperature changes of the air in the bottle. Repeated trials confirm that a predictable relationship exists between air temperature and changes in air pressure.]

Air pressure and temperature relationships

1. Compressing the air by squeezing the bottle was accompanied by a(n) [(*decrease*) (*increase*)] in the temperature of air inside the bottle.

2. The expansion of air that occurred when the bottle was allowed to return to its original shape and volume was accompanied by a(n) [(*decrease*)(*increase*)] in the temperature of air inside the bottle.

3. These observations indicate that when air is compressed, its temperature increases, and when air expands, its temperature [(*decreases*)(*increases*)].

4. Air pressure in the open atmosphere always decreases with an increase in altitude. This happens because air pressure is determined by the weight of the overlying air. Air rising through the atmosphere expands as the pressure acting on it lowers and, in turn, its temperature [(*decreases*)(*increases*)].

5. Air sinking in the atmosphere is compressed as the air pressure acting on it increases, and its temperature [(*increases*)(*decreases*)].

Making clouds appear and disappear

Now imagine removing the bottle cap and pouring a few milliliters of water into the container. Twist and turn the bottle to wet the inner surface before pouring out the extra water. Then, reseal with the cap. After a couple of minutes enough water will evaporate to saturate the bottle volume.

Next reopen the bottle to introduce some smoke to the air inside. The smoke is being added because atmospheric water vapor requires particles (nuclei) on which to condense. In the atmosphere, particles acting in the same way are called *cloud condensation nuclei*. [Place the bottle on its side, open the bottle, and push down to flatten the bottle to about half its normal diameter. Have another person light a match, blow it out, and insert the smoking end into the open bottle. Quickly release your pressure on the bottle so it returns to its original shape and the smoke from the extinguished match flows inside. Quickly cap the bottle tightly.]

Now apply and release pressure on the bottle as before, noting the temperature changes. When the bottle is allowed to spring back to its original shape, the temperature lowers—and a cloud appears in the bottle! The cloud is evidenced by a change in air visibility. Repeating the process of applying and releasing pressure will cause the cloud to appear and disappear.

6. The cloud forms when the pressure acting on the saturated air lowered and the temperature [(*increased*)(*decreased*)].

7. Most clouds in the atmosphere form in a similar way as the cloud in the bottle. With the temperature change due to expansion, some of the water vapor in the saturated air must [(*sublime*)(*evaporate*)(*condense*)], thereby forming cloud droplets.

8. Once you have a cloud in the bottle, squeeze the bottle to make the cloud disappear. The cloud disappears when the air temperature is raised by [(*compression*)(*expansion*)]. The change in temperature results in evaporation of the cloud droplets.

9. It can be inferred from this investigation that in the open atmosphere where it is cloudy, air is generally [(*rising*)(*sinking*)] and cooling. Where the atmosphere is clear, the air is generally moving in the opposite direction.

10. Generally, high pressure areas in the atmosphere tend to be clear because air in them experiences [(*upward*)(*downward*)] motion. Low pressure areas tend to have clouds because air in them experiences motion in the reverse direction.

Vertical motion, pressure change, and temperature change are of major importance in the formation and dissipation of most clouds. However, another major factor is at work. At any given temperature, there is a maximum concentration of water vapor that can ordinarily occur. This condition, called *saturation*, occurs when the temperature and dewpoint are equal. (The dewpoint is the temperature to which the air must be cooled at constant pressure to reach a relative humidity of 100%.) Air always contains some water vapor, but usually less than the maximum possible for its temperature. Cloud formation requires saturation be achieved so that, with further cooling, excess water vapor can change to the liquid (or solid) state. Thus, the atmosphere must undergo some process whereby saturation is achieved and further cooling takes place if clouds are to form.

The cloud-in-a-bottle investigation shows how atmospheric processes can produce saturation by changing air pressure. Lowering air pressure leads to lower air temperatures and, if enough water vapor is available, saturation is achieved.

As directed by your course instructor, complete this investigation by either:

1. *Going to the Current Weather Studies link on the course website, or*
2. *Continuing to the Applications section for this investigation that immediately follows in this Investigations Manual.*

Investigation 6A: Applications

CLOUDS, TEMPERATURE, AND AIR PRESSURE

Clouds are conglomerations of tiny water droplets (and/or ice particles) that formed from condensation (or deposition) of water vapor. This requires air to have a relative humidity of 100%, meaning saturation. Therefore, atmospheric processes that lead to saturation above Earth's surface form the clouds that are prevalent in the sky. The first part of this Investigation 6A demonstrated how an air parcel containing water vapor, when rising through the atmosphere, would expand and could eventually cool to the dewpoint.

Figure 1 is the surface weather map for 00Z 7 MAR 2010, local evening time. At map time generally clear or only partly cloudy conditions were prevalent over the eastern half of the U.S. where a fair weather High dominated. Two main weather systems were bringing clouds and areas of precipitation to regions of the country. One relatively weak low-pressure system with its accompanying warm and cold fronts was centered in the middle of the country. A second more complex set of low-pressure centers represented an expansive storm system that was impacting the West Coast. A long sequence of storm systems had been hitting the West coast from southern California to Washington State. While wintertime is the season where most precipitation falls along the coast, this pattern of storms bringing heavy precipitation is more closely linked to the El Niño conditions that existed in the central Pacific Ocean during the 2009-2010 winter season.

11. The portion of the western storm system most directly impacting the coast at map time was a low-pressure center located just west of Los Angeles with an occluded and then cold front curving near the coast before swinging southwestward into the Pacific. Radar echo shadings showed that there **[(_was_)(_was not_)]** precipitation occurring along the southern California coast.

12. The sky cover condition reported at Los Angeles, on the coast just east of the Low at the end of the occluded front, was **[(_clear_)(_partly cloudy_)(_overcast_)]**. The sky cover at San Diego to the south was obscured by the numerical label for the pressure system value, but was presumably the same.

13. The wind direction at San Diego (the wind arrow is seen below the pressure value box on the map) was from the **[(_north_)(_east_)(_west_)(_south_)]**.

14. This direction **[(_was_)(_was not_)]** consistent with the counterclockwise flow pattern around the Low associated with the frontal system.

15. With the front approaching the southern California coast, there would be **[(_frontal_)(_orographic_)]** lifting of the air along the frontal surface.

16. The air flow shown by the wind arrows at San Diego and Los Angeles, was along the

coast and partially <u>toward</u> the mountains to the east that parallel the coast forcing the air upward, resulting in [(***frontal***)(***orographic***)] lifting of the air. Thus lifting of the air had multiple causes.

17. The temperature at San Diego was reported as 58 °F and the dewpoint as 47 °F. With this difference between temperature and dewpoint, the air at the surface in San Diego [(***was***) (***was not***)] saturated.

18. **Figure 2** is the Stüve diagram from San Diego, California (NKX) rawinsonde observation at 00Z on 7 MAR 2010 *(100307/0000)*, the same time as the Figure 1 surface map. The temperature curve is the heavy plotted curve to the right and dewpoint is the curve to the left, respectively. The difference of temperature and dewpoint curves at the surface (lowest plotted level) tells us that the air at the ground in San Diego [(***was***)(***was not***)] saturated.

19. Air rising above the surface at San Diego was cooling due to expansion. The temperature became within about a degree of the dewpoint at about [(***1000***)(***900***)] mb. (Because of sensor behavior, temperatures may not always equal dewpoints on radiosonde profiles where air is saturated. However, meteorologists consider that when the values are within a few degrees saturation and clouds may exist in the area.)

20. This near equality of temperature and dewpont readings indicated that the air probably [(***was***)(***was not***)] saturated at that level. Based on this evidence we can assume the base of the cloud over San Diego to be at about that level.

21. Cloud conditions would exist where the temperature and dewpoint were reported as nearly equal. If they separate by several degrees at a higher level, we would assume the air [(***was still***)(***was no longer***)] saturated.

22. Taking the bottom of the cloud to be at the altitude where the temperature and dewpoint are nearly equal (900 mb), and the top of the cloud to be the altitude where they are separated by several degrees, the top of existing cloud conditions in the lower troposphere over San Diego would be near [(***800***)(***730***)(***700***)] mb.

23. From upper air map information (not shown), the 900 mb level occurred at about 1000 m. The pressure level at which temperatures and dewpoints began separating by several degrees occurred at about 2500 m. Therefore, the vertical extent of the cloud between these two altitudes over San Diego was about [(***1000***)(***1500***)(***2500***)] m.

Many of the cities along the West Coast are especially scenic because they are near the ocean but still within sight of the coastal mountains. The meteorological "price" to be paid for these topographical attractions is the influences they have on the weather. The moisture supplied by the ocean and the storm systems that come at some times of the year encounter the rising terrain. This often proves sufficient for frontal and orographic lifting of the air leading to precipitation. Investigation 6B will consider further details of rising air motions.

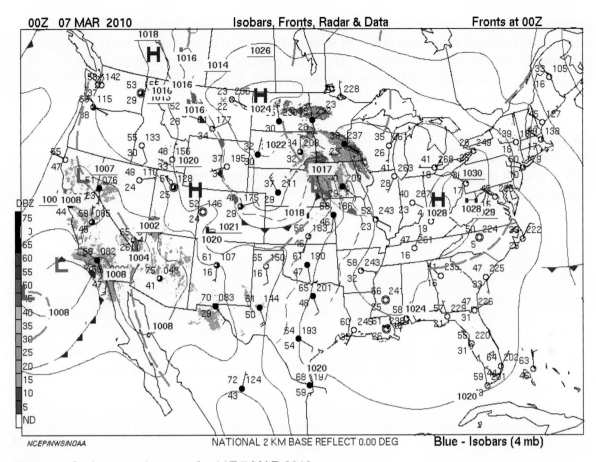

Figure 1. Surface weather map for 00Z 7 MAR 2010.

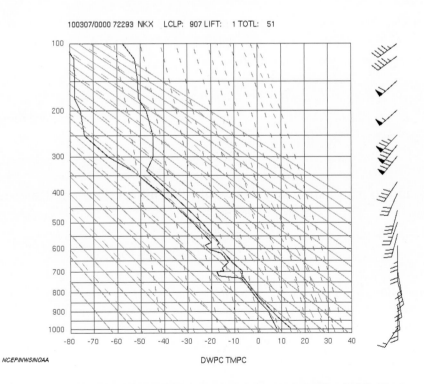

Figure 2. Stüve diagram for San Diego (NKX), CA sounding at 0000Z 7 MAR 2010.

<u>Suggestions for further activities:</u> You might compare Stüve diagrams for the station nearest you with surface cloud reports or with satellite observations to see ways whereby the existence of clouds can be determined. If you have periods of fog or predictions for it, you might check meteograms to follow the temperature and dewpoint values over time. (Fog is a cloud in contact with Earth's surface.)

Another process achieving saturation is the mixing of hot, humid air with cold, dry air in the formation of contrails (condensation trails - cloud-like streamers frequently observed to form behind aircraft flying in clear, cold air). This process is described at: *http://cimss.ssec.wisc .edu/wxwise/class/contrail.html.*

Objectives:

As air moves vertically in the atmosphere, it experiences changes in surrounding atmospheric pressure. These changes in pressure allow a rising parcel (a term used in meteorology to imply a small volume or body of air) to expand as surrounding pressures decrease, and cause a sinking parcel to be compressed as surrounding pressures increase. Rising, <u>unsaturated</u> air expands and cools at a rate of 9.8 C° per 1000 m (5.5 F° per 1000 ft). This is called the **dry adiabatic lapse rate**. Sinking unsaturated air warms at the same rate.

A rising unsaturated (relative humidity less than 100%) air parcel expands and cools, and may eventually become saturated. Upon saturation, further ascent will continue the expansional cooling. But, some heating will take place within the parcel because of condensation (or deposition at low temperatures). This warming occurs as latent heat is released to the air in the parcel when water vapor condenses into droplets or deposits as ice crystals. The simultaneous cooling by expansion and warming by condensation or deposition results in a **net** (observable) cooling rate, called the **saturated (or moist) adiabatic lapse rate**, that is lower than the cooling rate for unsaturated air. The saturated adiabatic lapse rate varies with temperature, but can be considered to average about 6 C° per 1000 meters (3.3 F° per 1000 ft).

After completing this investigation, you should be able to:

- Describe how to use a Stüve diagram to follow atmospheric temperatures and pressures.
- Determine the temperature of air that rises or sinks in the atmosphere.
- Describe how the water vapor saturation of air can affect atmospheric temperatures.

Introduction:

The **Figure 1** Stüve diagram in this investigation includes lines representing the adiabatic processes of dry (unsaturated) and saturated air. The solid, straight green lines sloping from lower right to upper left in the body of the chart graphically represent the dry adiabatic lapse rate, showing visually the temperature change of an unsaturated air parcel that is undergoing vertical motion in the atmosphere. The dashed, curved blue lines sloping from lower right to upper left represent the temperature change of saturated air undergoing vertical motion, the saturated adiabatic lapse rate. **Locate an air parcel with temperature 17 °C and pressure 1000 mb by placing a dot on the chart on the 1000 mb horizontal line where 17 °C would occur.**

1. If this air rises as <u>unsaturated</u> (dry) air from 1000 mb, determine its temperature at 500 mb by following the solid, straight green *dry adiabatic lapse rate* line passing through the starting point, up to 500 mb. At 500 mb, the temperature of the unsaturated air parcel is about **[(-5)(-35)(-45)]** °C.

2. If this air rises as <u>saturated</u> air from 1000 mb, determine its temperature at 500 mb by following the dashed, curved blue *saturated adiabatic lapse rate* line passing through the starting point, up to 500 mb. At 500 mb, the saturated air parcel's temperature is approximately **(*-15*)(*-25*)(*-35*)]** °C.

3. At 500 mb, the temperature of the unsaturated air parcel is **[(*lower than*)(*the same as*) (*higher than*)]** the temperature of the saturated air parcel.

4. This comparison demonstrates that rising unsaturated, clear air cools **[(*more*)(*less*)]** than rising saturated, cloudy air over the same interval.

5. Begin once again with unsaturated air at 17 °C and 1000 mb. Because it is unsaturated, its relative humidity initially is **[(*greater than*)(*equal to*)(*less than*)]** 100%.

6. As this air rises, assume it becomes saturated at 800 mb. Being unsaturated from 1000 mb to 800 mb, it will follow a **[(*dry*)(*saturated*)]** adiabatic lapse rate line.

7. Being saturated at 800 mb, its relative humidity is now **[(*greater than*)(*equal to*) (*less than*)]** 100%.

8. As the air continues to rise, it will follow a **[(*dry*)(*saturated*)]** adiabatic lapse rate line.

9. Continue the ascent to 500 mb. The air parcel temperature is now approximately **[(*-18*)(*-27*)(*-34*)]** °C.

10. This temperature is **[(*higher than*)(*equal to*)(*lower than*)]** the temperature achieved by the unsaturated parcel that ascended dry adiabatically the entire way to 500 mb in item 1.

11. If condensation was occurring during the ascent from 800 mb to 500 mb, the air parcel would have **[(*gained*)(*lost*)(*had no change in*)]** water vapor during the ascent.

12. Throughout this saturated portion of the ascent, the relative humidity of the air parcel is **[(*greater than 100%*)(*100%*)(*less than 100%*)]**.

13. Assume that all the water that condensed (or deposited) during the ascent was immediately lost as precipitation from the parcel. Therefore, if the air parcel at 500 mb begins to descend, it will experience warming by compression and immediately become an unsaturated parcel. As the parcel sinks back to the 1000-mb level, it will warm at the dry adiabatic lapse rate, as shown by following the dry adiabatic lapse rate line down from the point at 500 mb. When it arrives back at 1000 mb, its temperature is **[(*17*)(*27*) (*37*)]** °C.

14. This parcel's final temperature is **[(*higher than*)(*the same as*)(*lower than*)]** its beginning temperature when it was initially at 1000 mb.

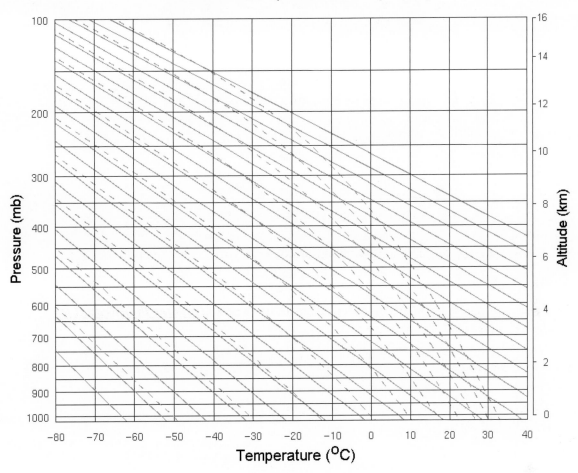

Figure 1. Vertical Atmospheric (Stüve) Chart with adiabats.

15. The relative humidity of this air parcel is now **[(*greater than*)(*equal to*)(*less than*)]** what it was when it began its journey at 1000 mb.

16. The change from the initial parcel temperature to the final parcel temperature at the 1000-mb level was caused by condensation (or deposition) which **[(*releases*)(*absorbs*)]** latent heat.

As directed by your course instructor, complete this investigation by either:

1. *Going to the Current Weather Studies link on the course website, or*
2. *Continuing to the Applications section for this investigation that immediately follows in this Investigations Manual.*

Investigation 6B: Applications

RISING AND SINKING AIR

The Investigations 6A **Applications** section examined a winter storm that brought precipitation to southern California. San Diego was receiving rain ahead of a landfalling frontal system. Here we will further consider where vertical air motions associated with the passage of a front were depicted on a Stüve diagram.

Figure 2 is the surface weather map for 00Z 10 MAR 2010, Tuesday evening. At that time a sprawling storm system was circulating across the central U.S. Heavy rains and some snow were affecting regions about the system. Following that system was another frontal system shown in two parts from eastern Idaho to northern Mexico that brought more scattered rain and snow to the Intermountain West. This pair of fronts was the system that had come ashore from the Pacific earlier and was examined in Investigations 6A **Applications**. And another frontal system was along the northern California to Washington state coasts ready to make its entrance. We will take a look at the atmospheric conditions over Salt Lake City, Utah that accompanied the frontal system there.

17. The surface temperature at Salt Lake City was 44 °F and the dewpoint was 35 °F. Therefore, the surface air at Salt Lake City [(*was*)(*was not*)] saturated.

18. Based on the station model sky cover and the radar echo shadings of precipitation shown around the Salt Lake City area, clouds [(*were*)(*were not*)] present somewhere above the surface.

19. Therefore, we can be confident that saturated air [(*was*)(*was not*)] present above the surface over the Salt Lake City area.

20. Salt Lake City in the Great Basin region is surrounded by mountains so wind patterns would often provide orographic lifting of the air in the area. Also, the Figure 3 weather map shows a [(*high-pressure system*)(*cold front*)] had just passed Salt Lake City. Both of these mechanisms were providing lifting of the air over Salt Lake City.

21. With both lifting mechanisms for the air over the Salt Lake City area, we can be reasonably sure that [(*rising*)(*sinking*)] vertical motions occurred above the surface.

Figure 3 is the Stüve diagram from the Salt Lake City (SLC) rawinsonde observation at 0000Z 10 MAR 2010, the same time as the Figure 2 surface map. Note that the lowest pressure level shown for the temperature and dewpoint profiles in Salt Lake City is at about 860 mb. This is the *station* pressure, the air pressure measured by a barometer at the land's surface. The surface elevation of the rawinsonde station in Salt Lake City is 1289 meters (4228 feet). The Figure 2 map's decoded surface air pressure, 1004.0 mb, is the *sea-level* pressure reading, the station pressure corrected to what it would be at sea level!

22. The temperature and dewpoint profiles from the surface in Salt Lake City up to about 800 mb show that the air **[(*was*)(*was not*)]** saturated.

23. The temperatures and dewpoints were approximately equal from about 800 mb to about 640 mb. This near equality means it **[(*was*)(*was not*)]** likely that cloud conditions were present in that layer.

24. From about 860 mb to 800 mb, the temperature profile was parallel to the adjacent **[(*straight, solid, green dry*)(*curved, dashed, blue saturated*)]** adiabatic lapse rate line printed on the diagram.

25. From about 800 mb to about 640 mb, the temperature and dewpoint profiles curved approximately "parallel" to the adjacent **[(*straight, solid, green dry*)(*curved, dashed, blue saturated*)]** adiabatic lapse rate line printed on the diagram.

26. The following values come from the rawinsonde text data (not shown). Over Salt Lake City, 861 mb (surface) occurred at 1289 m where the temperature was 6.6 °C and 793 mb was at 1953 m where the temperature was 0 °C. Therefore, the temperature difference between those levels was 6.6 C° over an altitude change of 669 m. The temperature lapse rate was therefore an equivalent 9.87 C° per kilometer. This lapse rate value **[(*was*)(*was not*)]** essentially the same as the theoretical 9.86 C° per kilometer. Unsaturated rising air really does follow an adiabatic process!

27. Additionally, at the top the identified cloud layer over Salt Lake City the temperature was -11.1 °C at 639 mb (3658 m). The temperature difference between the 793-mb cloud base and the 639-mb cloud top was 11.1 C° over an altitude change of 1705 m. The temperature lapse rate was therefore an equivalent 6.5 C° per kilometer. This value **[(*was*)(*was not*)]** reasonably consistent with the average 6 C° per kilometer for clouds. Saturated rising air also really does follow an adiabatic process!

Stüve diagrams of actual observations confirm that vertical atmospheric motions do follow the theories! Call up these Stüves as dramatic weather changes affect your area. Weather systems (Highs, Lows, fronts) force air to move vertically causing accompanying temperature changes. Flow forced over higher and lower elevations, particularly in the western U.S. also drives atmospheric temperature patterns.

Suggestions for further activities: You might print out the text data of rawinsonde observations and plot them on a blank Stüve diagram (available on the website) when Highs, Lows, and fronts pass nearby. Then compare local cloud and sky conditions with the temperature and dewpoint profiles you have plotted.

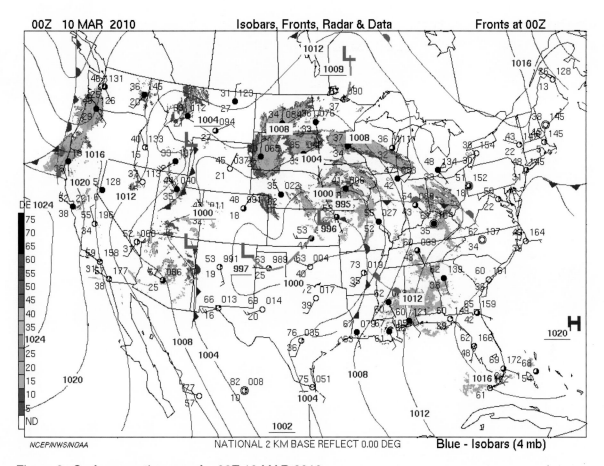

Figure 2. Surface weather map for 00Z 10 MAR 2010.

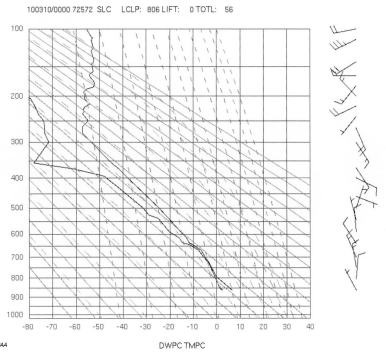

Figure 3. Stüve diagram for Salt Lake City (SLC), UT for 0000Z 10 MAR 2010.

Objectives:

Rain and snow are not random and capricious acts of nature. This is especially evident when weather-radar echo patterns signifying precipitation are viewed along with weather maps and satellite images for the same times.

The basic mechanism for the formation of clouds and precipitation is the uplift and consequent cooling of air by expansion. In fact, uplift of air along the sloping surface of a front is the principal mechanism whereby the circulation in lows produces clouds and precipitation. Clouds and precipitation also may be associated with the upward branch of a convection current, uplift of air along the windward slopes of a mountain range, or convergence of surface winds. One of the most useful tools in following the development and movement of areas of precipitation is weather radar.

After completing this investigation, you should be able to:

- Describe different mechanisms leading to the formation of clouds and precipitation in low pressure systems.
- Locate areas of precipitation based on weather radar depictions.
- Indicate the general relationship between the uplift of air and the formation of clouds and precipitation.

Introduction:

1. Recall that a *front* is a line drawn on a surface weather map that marks a narrow transition zone between air masses differing in density (due to contrasts in temperature and/or humidity). An *air mass*, in turn, is a huge volume of air covering perhaps tens of thousands of square kilometers having generally uniform temperature and humidity characteristics in its horizontal extent.

   A mass of cold, dry air is [(***denser than***)(***not as dense as***)] a mass of warm, humid air. Consequently, warmer (lighter) air is forced to rise above the sloping frontal surface overlying colder (denser) air.

2. Ascending unsaturated (clear) air [(***expands***)(***is compressed***)], cooling at about 10 Celsius degrees per 1000 meters of ascent (5.5 Fahrenheit degrees per 1000 feet).

3. The warmer air rising above the frontal surface expands and cools as it ascends, and its relative humidity [(***increases***)(***decreases***)]. If saturation is achieved, clouds develop and from those clouds, rain or snow may fall.

4. The relative humidity of saturated (cloudy) air is **[(_50%_)(_98%_)(_100%_)(_105%_)]**.

5. Clouds (and perhaps precipitation) can develop in the ascending branch of a convection current, along a front, and up the windward slopes of a mountain range. The ascending branch of a convective current may produce an upwardly billowing cloud known as a **[(_cumulus_)(_stratus_)(_cirrus_)]** cloud.

6. Prevailing winds blow from west to east across over most of North America. Winds that blow onshore from the Pacific Ocean are forced up the windward slopes of the Cascade Mountain Range in the Pacific Northwest. Hence, the heaviest precipitation falls on the **[(_western_)(_eastern_)]** slopes of the Cascades.

As directed by your course instructor, complete this investigation by either:

  1. *Going to the Current Weather Studies link on the course website, or*
  2. *Continuing to the Applications section for this investigation that immediately follows in this Investigations Manual.*

Investigation 7A: Applications

PRECIPITATION PATTERNS

Over the weekend prior to Monday, 22 March 2010, a storm system formed over the southern Plains states and swirled its way into the Southeast. The mix of cold and warm air streams brought large snowfall accumulations to the southern Plains while rain occurred elsewhere. Some thunderstorms in the Southeast became severe and produced considerable wind damage.

Figure 1 is the surface weather map for Monday morning, 11Z 22 MAR 2010. The storm was centered over the central Tennessee-northern Alabama-northwestern Georgia area as denoted by the semi-circular area of light precipitation shown by radar echoes. Shadings on the map denoted where a national network of NWS radars was detecting targets, typically precipitation (rain and/or snow). Light blues and greens on-screen were typical of light rain or snow while yellow patches with embedded red speckles generally signified heavier showers, and thunderstorms like those in eastern North Carolina and across Ohio. The radar shading scale was shown along the left edge of the map area. A larger arc of heavier precipitation curved around that central area and extended northeastward into New England. Further patches of precipitation were detected by radar along the Atlantic Coast southward to the Florida peninsula and across Cuba.

7. A dashed orange arc curved along the Ohio River from southern Illinois to West Virginia. This line denoted a pressure trough or region of lower pressure which followed the curves of the neighboring isobar lines. In West Virginia an *L* marked the center of lowest pressure. From near the L a short occluded front (purple with triangles and semicircles on the same side) was shown to North Carolina. In North Carolina, a *triple point* existed where the occluded front connected to a short **[(*warm*)(*cold*)(*stationary*)]** front which extended to the Atlantic coast at the North Carolina-Virginia border.

8. From the triple point, a **[(*warm*)(*cold*)(*stationary*)]** front curved generally southeastward from South Carolina out over the Atlantic along the coast and across the island of Cuba.

9. Further, stretching from eastern Ohio to Massachusetts and along the northern portion of the Atlantic coast, was a **[(*warm*)(*cold*)(*stationary*)]** front.

10. The winds at New York City just south of the front there and at Cape Hatteras, North Carolina south of that front were generally from the **[(*south or southeast*)(*east or northeast*)(*west or northwest*)]**. These wind directions were flowing from the ocean delivering humid air up over the nearby frontal surfaces in a pattern called *overrunning*.

11. Triangular symbols along the front from South Carolina to Cuba showed the front in this region was advancing generally toward the **[(*south*)(*east*)(*north*)(*west*)]**.

12. Comparing the temperature (75 °F) and dewpoint (75 °F) at Nassau, Bahamas east of the front, with the temperature (65 °F) and dewpoint (60 °F) at Miami, Florida west of the front, indicates that the air was [(***warmer and more humid***)(***cooler and less humid***)] east of (ahead of) the cold front.

13. The radar echoes displayed along all of these map features suggested that there would have been [(***frontal***)(***orographic***)] lifting mechanisms for the air.

14. Besides the lifting of the air, water vapor must be present for condensation and droplet growth leading to precipitation. The wind directions at New York City and Cape Hatteras suggested that the primary water vapor supply was moisture evaporated from the nearby [(***Pacific Ocean***)(***Atlantic Ocean***)(***Gulf of Mexico***)]. When the storm was forming earlier and was located in the southern Plains, the moisture brought into the circulation at that time was likely from the Gulf.

Figure 2 is the water vapor satellite image for Monday morning, 1115Z 22 MAR 2010, approximately the same time as the Figure 1 map. Water vapor imagery detects the amount of water substance between about 700 mb and 400 mb (10,000 to 25,000 feet) in the middle troposphere. Bright white shadings denote high clouds while medium gray shades are associated with extensive water vapor amounts. Dark areas are relatively dry air.

15. The water vapor satellite image indicates that there [(***was***)(***was not***)] extensive high cloudiness where the Figure 1 map's radar echoes showed relatively heavy precipitation was occurring.

16. The dark ribbon that curves from southern West Virginia across western North Carolina, central South Carolina, eastern Georgia and northern Florida [(***did***)(***did not***)] generally coincide with the area on the Figure 1 surface map where skies were partly cloudy and precipitation was not shown.

17. Knowing the presence of a surface Low in the region at the time, the swirl shown in the water vapor pattern of lighter gray shadings over the Southeastern U.S. suggests that the middle tropospheric circulation associated with the storm system was [(***clockwise***)(***counterclockwise***)] as seen from above.

Elsewhere on the Figure 2 water vapor image, there is evidence of another major counterclockwise swirl centered over eastern Montana. This identified another cyclonic storm that was developing at the time and was advancing into the north-central U.S. This was shown on the Figure 1 surface map by scattered precipitation resulting from combined frontal and orographic lifting in that region. While that system is not clearly evident from surface map conditions, the mid tropospheric comma-shaped swirl in the water vapor image displayed it well.

Suggestions for further activities: When precipitation is expected or occurring in your area, you can consult local weathercasts for observations of the location and total amounts over a

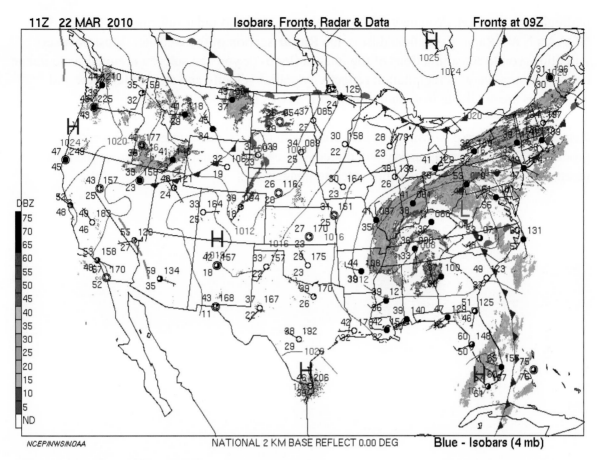

Figure 1. Surface weather map for 11Z 22 MAR 2010.

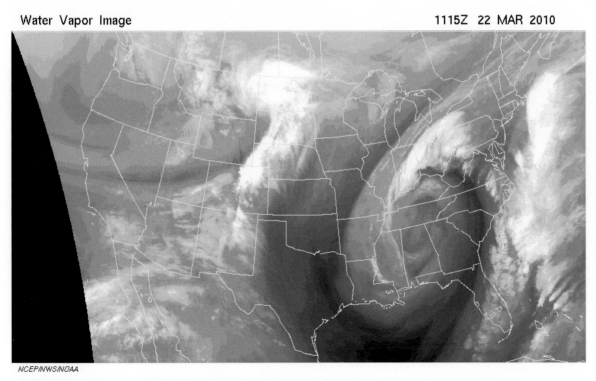

Figure 2. Water vapor satellite image for 1115Z 22 MAR 2010.

particular time period. You may wish to explore the current radar information provided by NOAA's NWS Radar page, linked from the **Radar** section of the website. Most browsers will allow you to run animations of the images (loop) or you can select a regional scale image below the large view. Precipitation reflectivity, radial velocity and storm total view and animations are available from individual stations by clicking on the map.

Remember, the course website provides both static images and animations of weather products. There, you can quickly and directly observe thumbnail animations of the most recent radar and IR and water vapor satellite imagery.

Investigation 7B:

DOPPLER RADAR

Objectives:

Weather radar has routinely provided valuable information on storm size, shape, intensity, and direction of movement for many years. With the advent of Doppler radar, it has also become possible to determine details of atmospheric circulation within a storm. For instance, air motions that indicate possible tornado development can be detected. This information provides the public with advance warning of severe weather and saves lives.

After completing this investigation, you should be able to:

- Describe aspects of the actual wind that are detected by Doppler radar.
- Determine the speed of the wind towards or away from the radar site.
- Construct the wind pattern as detected by Doppler radar.

Materials: Red and green pens or pencils.

Introduction:

Radar detection of motion is based on the *Doppler effect*, the change in frequency (or phase) of a sound or electromagnetic wave reaching a receiver when the receiver and source are moving relative to one another. A frequency shift occurs when a radar signal is reflected from a moving target such as a cluster of raindrops. If the raindrops are moving toward the radar, the reflected signals returned to the radar have a higher frequency than if the target were stationary. On the other hand, if the raindrops are moving away from the radar, the returned signal's frequency is lowered. The magnitude of the frequency shift is a measure of the parts of raindrops' motions that are directly toward or directly away from the radar.

Doppler weather radar is especially useful for detection of severe weather conditions. One of the most devastating and potentially deadly of severe weather phenomena is the tornado. A tornado is a rapidly rotating column of air in contact with the ground. Tornadoes are almost always associated with thunderstorms.

Before tornadoes develop their intense, ground-level circulation, a broader-scale horizontal rotation is often evident within the parent thunderstorm. This internal thunderstorm rotation is called a *mesocyclone*. As the air entering the thunderstorm begins to swirl in the mesocyclone, raindrops are carried along and reflect radar energy back toward the radar antenna.

1. **Figure 1a** is a schematic view of a radar beam detecting a mesocyclone, depicted as a rotating cylinder embedded within a severe thunderstorm. As viewed from above in the Northern Hemisphere, the mesocyclone's rotation is typically counterclockwise.

Figure 1b is a view from above with the size of the mesocyclone exaggerated (not to scale). The dashed lines represent the radar beam in positions 1 through 5 as it sweeps through the mesocyclone. The arrows drawn around the mesocyclone column in Figure 1b represent the actual wind at dots located at the tails of the arrows. Each arrow shows the instantaneous direction of air movement at that point, and the length of the arrow represents the speed of that wind.

In this example, the actual winds circulating around the mesocyclone all have the same speed, as shown by arrows whose lengths are [(*the same*)(*different*)].

2. From one location to another around the mesocyclone, the wind directions are [(*the same*)(*different*)].

3. Doppler radar detects only those motions or components of motion that are directly toward or away from the radar. At two locations within the rotating wind pattern no air motion is detected by the Doppler radar beam. At those points, the actual wind blows along paths perpendicular to the radar beam (dashed line). Hence, the Doppler radar senses no Doppler wind speed at these locations. These two locations are sensed by the radar when its beam is in the [(*1*)(*2*)(*3*)(*4*)(*5*)] position. **Draw a small circle around each of the dots associated with those two arrows to denote this zero Doppler wind speed.**

4. Two arrows on the column circle are along the direction of the radar beam, one directly towards the radar and one directly away. These two locations are sensed when the radar beam is at the [(*1 and 3*)(*4 and 2*)(*3 and 5*)(*5 and 1*)] positions. Because these arrows are oriented directly away from or directly toward the radar site along the beam direction, the radar will sense the full wind speed, away or toward, respectively.

   Where the wind arrow is oriented directly toward the radar, use a green pencil to trace a bold, green arrow of the same length and direction atop the wind arrow. Where the wind arrow is oriented directly away from the radar, use a red pencil to trace a bold, red arrow of the same length and direction atop the wind arrow. [The NWS color convention uses "cool" colors such as greens and blues for motions toward the radar and "warm" colors such as reds and oranges for those away from the radar.]

5. At the other four arrow locations shown around the mesocyclone, wind arrows are neither directly toward or away, nor perpendicular to the radar beam direction. Where the radar beam direction and the actual wind arrow make an angle other than 0 or 90 degrees, Doppler radar senses only the component of the total motion that is directly toward or away from the radar. **For the two arrows that are directed partly toward the radar, use the green pencil to draw approximately half-length green arrows, from the location dots, that are aimed directly toward the radar along the dashed beam direction.** These two locations are at the [(*2*)(*3*)(*4*)(*5*)] radar beam position.

6. **For the two arrows that are directed partly away from the radar, use the red pencil to make similar half-length red arrows, drawn from the location dots, which are**

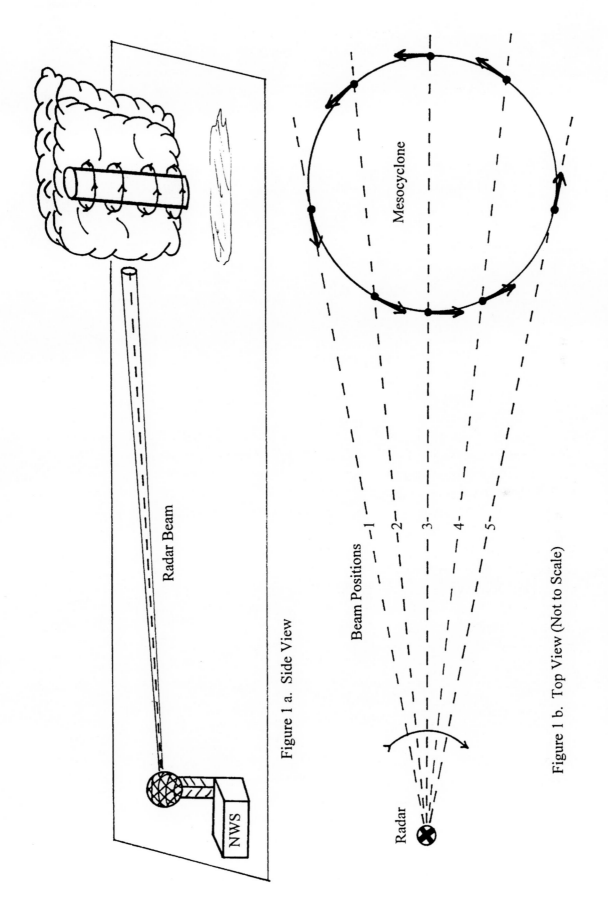

Radar Beam

Figure 1 a. Side View

Mesocyclone

Beam Positions

Radar

Figure 1 b. Top View (Not to Scale)

Figure 1. Schematic view of (a) radar beam detecting a mesocyclone with tornado and (b) sensing radial velocities.

aimed directly away from the radar. These two locations are at the **[(*1*)(*2*)(*3*)(*4*)]** radar beam position.

Finally, with the green pencil, shade across the semi-circular area of the mesocyclone where the arrows are green. Shade the lightest from near the zero position, becoming darker where the green arrow is longest. With the red pencil, shade across the portion of the mesocyclone where the arrows are red. Graduate the shading from lightest near the zero position, becoming darker where the red arrow is longest.

Observe the colored arrows of your mesocyclone depiction. These are the Doppler winds as detected by the radar utilizing the Doppler effect. Respond to Items 7 -11 as either **T** for true, or **F** for false, based on the colored arrow pattern of the radar display associated with the mesocyclone.

7. **[(*T*)(*F*)]**   The green arrows are directed toward the radar.

8. **[(*T*)(*F*)]**   The red arrows are directed away from the radar.

9. **[(*T*)(*F*)]**   Along the radar beam, when in position 3, the Doppler wind speed is zero.

10. **[(*T*)(*F*)]**   The green shaded area depicts air motions towards the radar.

11. **[(*T*)(*F*)]**   The red shaded area depicts air motions away from the radar.

The color scheme you have drawn on Figure 1b could represent the severe weather "signature" of a mesocyclone on a Doppler weather radar display. The signature has regions of green and red appearing on opposite sides of a radial line from the radar's location. Meteorologists have identified other identifiable Doppler radar patterns associated with fronts, gust fronts and outflow boundaries from thunderstorms, wind shear, and other forms of severe weather.

As directed by your course instructor, complete this investigation by either:

1. *Going to the Current Weather Studies link on the course website, or*
2. *Continuing to the Applications section for this investigation that immediately follows in this Investigations Manual.*

Investigation 7B: Applications

DOPPLER RADAR

On a several-day period centering on 24 March 2010, a significant spring-time weather situation occurred on the plains east of the Rocky Mountains. A surface Low formed in southeastern Colorado, with the area of the Front Range of the Rockies around Denver receiving six to twelve inches of snow by the evening of 23 March. The system drifted southeastward before heading to the northeast over the following couple of days. This slow movement brought abundant moisture into the storm's circulation and produced heavy rain and severe weather later in the storm's life.

12. **Figure 2** is the national surface weather map for 11Z 24 MAR 2010. Precipitation as shown by radar shadings [(***did***)(***did not***)] cover the area from eastern Colorado to Iowa and southward to eastern New Mexico and Texas at map time.

13. Pueblo, located in southeastern Colorado and indicated by a brown (on-screen) square, was situated northwest of the low-pressure center of the storm system marked by an L along the Oklahoma-Texas border. Based on the Denver wind direction (just north of Pueblo), winds across eastern Colorado were probably generally from the [(***north-northeast***)(***south-southwest***)].

14. The wind directions shown at Denver, Amarillo and Midland across west Texas [(***were***)(***were not***)] consistent with the counterclockwise circulation around the "backside" of a Low.

15. **Figure 3** is the Southwest regional surface map showing station models for many more stations than can be displayed on the U.S. map. The time of the regional map was 11Z 24 MAR 2010, the same time as the Figure 2 surface map. [Regional maps of the latest conditions can be accessed from the website **Surface** section.] The station model for Pueblo in east-central Colorado was marked by an X in the station circle indicating that the sky was obscured and the cloud coverage could not be determined. Pueblo's temperature was 29 °F, dewpoint 27 °F, and the present weather was two stars meaning light snow. (It was the snow that obscured the cloud cover observation.) The wind at Pueblo was generally <u>from</u> the [(***north***)(***east***)(***south***)(***west***)] at about 20 knots. This direction of surface wind flow was consistent across eastern Colorado.

16. **Figure 4** is the Pueblo, CO National Weather Service (PUX) Doppler radar displays of the Base Reflectivity on the left and the Base (radial) Velocity on the right at 1117Z on 24 MAR 2010, near the time of the Figure 2 and Figure 3 surface maps. The location of the radar site northeast of Pueblo is denoted by a black dot in the centers of the reflectivity and velocity displays. The reflectivity is related to the intensity of radar return signal. The reflectivity display pattern at Pueblo is shown keyed to the scale along the right side. The blue and green shadings (on-screen) widely spread over the area indicates a return

of radar energy to the receiver from generally light precipitation. The precipitation type denoted on the Figure 3 regional map was [(*__light rain__*)(*__light snow__*)(*__fog__*)].

17. Comparing the Pueblo reflectivity shading pattern in the left side of Figure 4 to the region denoting precipitation by radar shadings on the Figure 2 surface map showed that they [(*__generally agreed__*)(*__were widely displaced__*)] in maximum intensity and coverage location near the site.

In the Doppler Base (radial) Velocity display on the right in Figure 4, green hues indicate winds with radial components *toward* the radar and red hues denotes radial components of the wind *away from* the radar sited at the black dot. The scale at the lower right shows directions and magnitudes of the radial velocity in knots, negative values toward the radar while positive values are away. The incomplete magenta "oval ring" on the periphery of the display area indicates where the radar detects precipitation but cannot distinguish between wind motions toward or away from the site.

18. The green and red radial velocity shadings [(*__do__*)(*__do not__*)] generally cover most of the areas where the reflectivity return shadings of the left view indicated precipitation was occurring near the radar site.

19. In the Base Velocity view (right), the lighter shade boundary between reddish (on screen) and greenish shadings oriented generally west-northwest / east-southeast indicates 0 "Doppler wind speed". Draw a short straight line segment of "best fit" through the black dot to show the position of 0 Doppler wind speed near the radar site. This "zero-speed" situation occurs when the radar beam direction is [(*__perpendicular__*)(*__parallel__*)] to the actual wind direction (or there are calm conditions) and there is no wind motion component directly towards or away from the radar.

20. In the Base Velocity display, the main area of winds flowing <u>toward</u> the radar site (brightest green shading) is located generally to the [(*__north__*)(*__south__*)] of the radar site.

21. In the Base Velocity display, the main area of winds flowing <u>away from</u> the radar site (brightest red shading) is located generally to the [(*__north__*)(*__south__*)] of the radar site.

22. Draw a short (about 1 cm) arrow perpendicular to the 0 Doppler wind speed boundary (drawn in Item 8) from green to red shadings through the black radar site dot. Place an arrowhead on the red end to indicate the Doppler-detected wind direction at the station. The direction of your arrow signifying the wind direction in the lower layers of the atmosphere sensed by the radar signal [(*__is__*)(*__is not__*)] generally along the same direction as the near-surface winds reported in Item 4.

As the radar beam travels outward from the radar site, the beam curves downward less than the curvature of the underlying Earth's surface. Therefore, the beam is sampling air at increasing altitudes with distance outward from the radar site and shading patterns give information on wind speeds and directions at higher and higher altitudes.

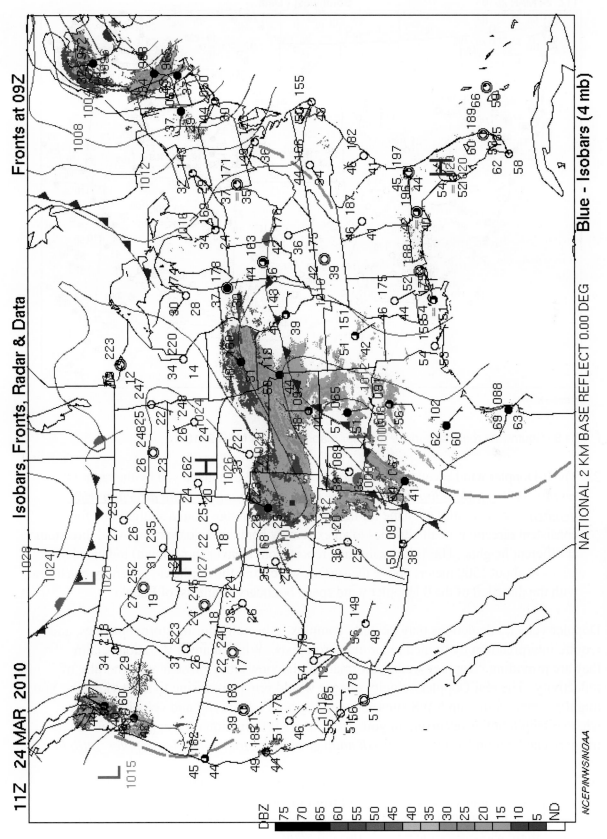

Figure 2. Surface weather map for 11Z 24 MAR 2010.

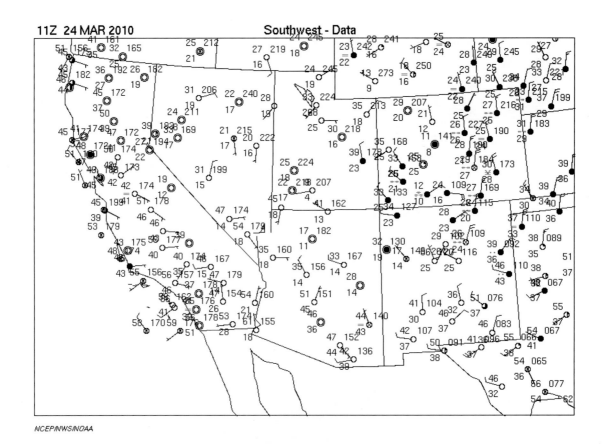

Figure 3. Southwest regional surface weather map for 11Z 24 MAR 2010.

23. The 0-Doppler wind speed boundary across the central area displayed by the radar is essentially [(*a straight line*)(*a curved line indicating changing wind direction with height*)]. The extremes of shape for this boundary can range from a straight line for consistent directions at all altitudes to an extremely bent curve for varying wind directions at different heights. The 12Z rawinsonde report from Denver (not shown) indicated that winds at about 1500 meters above the surface were from the northeast, generally consistent with the direction of the 0 Doppler wind speed boundary southeast of Pueblo.

Doppler velocity depictions may be quite complex, especially during storm episodes, and require interpretation by trained radar meteorologists. While many TV stations claim they are providing "Doppler" radar information, the views they present are generally of reflectivity. The real Doppler velocity images would prove confusing to most viewers. You might practice calling up NWS sites and viewing both reflectivity and velocity displays where precipitation is occurring in your area. More information on Doppler radar and its imagery can be found at *http://www.srh.noaa.gov/srh/jetstream/doppler/doppler_intro.htm*.

We will examine additional radar views of air motions in a later investigation dealing with tornadoes. Such radar views can be found from the website **Radar** section with the *NWS Radar Page* link. Additional discussion of radar imagery interpretation can be found at: *http://ww2010.atmos.uiuc.edu/(Gh)/guides/rs/rad/home.rxml*.

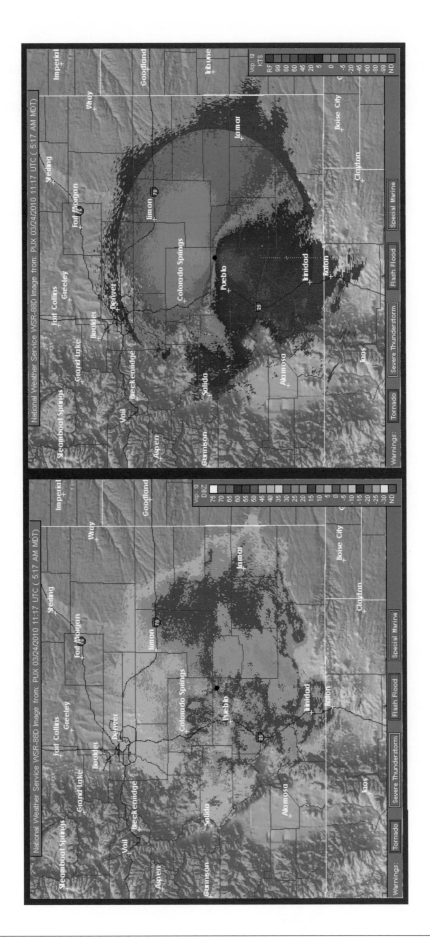

Figure 4. Pueblo, CO NWS Doppler radar display of Base Reflectivity (left) and Base (radial) Velocity (right) at 1117Z 24 MAR 2010.

<u>Suggestions for further activities:</u> The radar images in this investigation were from sites located via "NWS Radar Page" link from the course website. Another source of radar imagery is *http://www.intellicast.com/*. Also, a discussion of Doppler technology, including storm relative velocities, can be found at: *http://www.crh.noaa.gov/lmk/soo/88d/index.php* with additional images at: *http://www.crh.noaa.gov/lmk/soo/88dimg/index.php*.

SURFACE WEATHER MAPS AND FORCES

Objectives:

Although the atmosphere is almost entirely a gaseous fluid, it is a system with physical mass that responds to gravity and other forces such as those arising from pressure differences over distance (gradients). Gravity holds the atmospheric shell to Earth as a thin layer over the solid and liquid surfaces of the planet. Frictional coupling with the planetary surface causes the atmosphere to rotate with the planet. By isolating the forces that act on a parcel of air, we can explain observed air motions and the various scales of atmospheric circulation.

After completing this investigation, you should be able to:

- Describe the horizontal forces that act on air parcels.
- Show the directions toward which these atmospheric forces act.
- Relate these horizontal forces to the winds reported on weather maps.

Materials: Two "3x5" cards (or two cards 3 inches by 5 inches cut from stiff paper), scissors, tape, and pen or pencil.

Introduction:

Pressure Gradient Force

An air pressure gradient exists wherever air pressure varies from one place to another. This change in pressure over distance results in a force that puts air into motion.

1. The diagram to the right represents a portion of a surface weather map on which are plotted three straight, parallel isobars. Pressure is in mb units, and isobars are uniformly spaced and drawn with a 4-mb interval. **[(*High*)(*Low*)]** pressure is located across the top of the diagram.

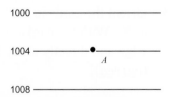

2. The diagram shows a pattern of air pressure changing over distance. Assuming that the atmosphere is initially calm, the only force acting horizontally on a parcel of air represented on the diagram at Point A is a pressure gradient force. **Draw an arrow about a centimeter in length starting at Point A and aimed directly towards the top in the diagram that depicts the direction the pressure gradient force would act.** Your arrow shows the pressure gradient force acting directly towards **[(*highest*)(*lowest*)]** pressure. This force is directed perpendicular to the isobar lines. The horizontal pressure gradient has given rise to a force that causes the air parcel at A to begin moving in the direction towards which the force is acting.

Coriolis Effect

Everywhere on Earth, except at the equator, objects moving freely across Earth's surface travel along curved paths. This turning is produced by Earth's rotation and is called the **Coriolis Effect**. The following demonstrates the impact of Earth's rotation on horizontally-moving objects.

Directions: First construct a rotating card device with two 3x5 file cards (or two cards 3 inches by 5 inches made from stiff paper). Following the diagram below, (i) cut an approximately two and one-half inch straight slit down the middle of one card (A), and (ii) cut a slit about one and one-half inch long in the middle of the other card (B). Fit the cards together as shown (iii), and lay them flat on the desk or table in front of you. Tape card (A) to the table with the long slit as shown (tape is dotted rectangles). Bend up the lower left and right corners of the loose card (B) to use as tabs. Pull the loose card horizontally towards you until the ends of the cuts meet to form a point of rotation. Be sure the moving card (B) can turn clockwise and counterclockwise around the point of contact. Make an **X** to mark the spot around which the card rotates.

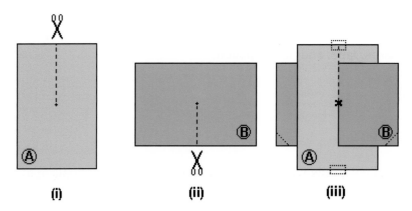

| (i) | (ii) | (iii) |

3. Orient the cards in the "cross" position as shown in the drawing. Place your pencil point at **X**. With the cards motionless, carefully draw a line <u>on the loose card</u> (B) along the cut-edge and directly away from you. The line you drew represents a path that is [(***straight***) (***curved***)].

4. Now investigate how rotation affects the path of your pencil line. Again, begin with the cards in the "cross" position and your pencil point at **X**. As you slowly pull the lower left tab of the loose card (B) towards you, slowly move your pencil point away from you along the cut-edge while drawing its path <u>on (B)</u>. The loose card is rotating counterclockwise as you do this. The line you drew is [(***straight***)(***curved***)].

5. You actually moved the pencil point along a path that was both straight *and* curved at the same time! This is possible because motion is measured relative to a frame of reference. In this investigation, there are two different frames of reference; one fixed and the other rotating. When the pencil-point motion was observed relative to the fixed card (A) and its cut-edge, its path was [(***straight***)(***curved***)].

6. When the pencil motion was measured relative to the rotating card (B), its path was **[(*straight*)(*curved*)]**. This apparent deflection of motion from a straight line in a rotating system is called the **Coriolis Effect** for Gaspard Gustave de Coriolis (1792-1843), who first explained it mathematically. Because Earth is a rotating system, objects moving freely across its surface, except at the equator, exhibit curved paths. This includes air parcels moving horizontally.

7. Now imagine yourself far above the North Pole and looking down on the Earth below. Think of the loose card (B) as being part of Earth's surface and that X represents the North Pole. From this perspective, Earth appears to rotate counterclockwise. You can observe the pencil point's motion relative to the Earth's surface (B). You see that as the pencil point moves along the cut-edge and away from the **X**, it draws a path on the rotating surface that **[(*is straight*)(*curves to the right*)(*curves to the left*)]**.

8. Now imagine yourself far above the South Pole and looking down on the Earth below. Again, think of the loose card (B) as being part of the Earth's surface and that X represents the South Pole. From this perspective, Earth appears to rotate clockwise. Rotating the loose card clockwise by pulling on the lower-right tab, you can observe that as the pencil point moves along the cut-edge and away from the **X**, it draws a path on the rotating card that **[(*is straight*)(*curves to the right*)(*curves to the left*)]**.

9. The effect of Earth's rotation on the path of objects moving across its surface is greatest at the poles, and diminishes to zero at the Equator. In summary, the Coriolis Effect causes objects freely moving horizontally over the Earth's surface in the Northern Hemisphere to appear to curve to the **[(*right*)(*left*)]**.

10. The Coriolis Effect causes objects in the Southern Hemisphere to appear to curve to the **[(*right*)(*left*)]**.

When investigating atmospheric motions, it is informative to analyze the forces acting on the air. However, the rotating-card activity you just completed shows that the observed curved motions are due to a rotating frame of reference and not due to a force. Consequently, an imaginary *Coriolis "force"* is invented to be applied along with real forces to describe motions of objects. The Coriolis force producing such curved motion is defined as always acting perpendicular to the direction of motion, to the right in the Northern Hemisphere to explain rightward turning, and to the left in the Southern Hemisphere to describe leftward turning.

Pressure Gradient Force, Coriolis Effect, Friction, and Weather Maps

11. On the following weather map segment, consider an air parcel at rest at Point *A*. An initial horizontal pressure gradient force **[(*is*)(*is not*)]** acting on the parcel.

12. Once horizontal motion has begun at this Northern Hemisphere location, the air parcel's path will be deflected to the **[(*right*)(*left*)]**.

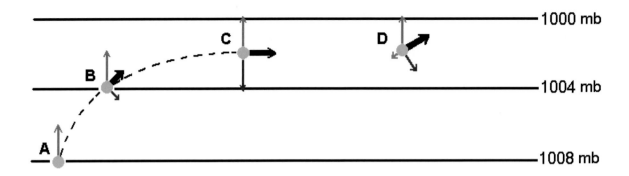

13. The moving air parcel follows the dashed curved path shown on the map. The thick black arrow at Point *B* shows the parcel's direction of motion at that location. At that instant, the longer thin red arrow represents the **[(*Coriolis*)(*Pressure Gradient*) (*Frictional*)]** force.

14. The shorter thin blue arrow represents the **[(*Coriolis*)(*Pressure Gradient*)(*Frictional*)]** force, which is acting at a right angle and to the right of the direction of motion.

15. As the air parcel speeds up, the Coriolis Effect increases and the parcel's motion continues to be deflected to its right. This continues until the parcel reaches Point *C* where the magnitude of the Coriolis Effect finally equals that of the pressure gradient force (which continues to act toward lowest pressure). At Point *C* the Coriolis Effect will be acting directly opposite to the pressure gradient force. The two forces are in balance. From Point *C* and onward, the air parcel will flow **[(*perpendicular*)(*parallel*)]** to the isobars. This flow is known as the **geostrophic wind**.

16. Point *D* shows the effect of friction on moving air. The **force of friction**, represented by the smallest green arrow drawn from Point *D*, always acts opposite to the direction of motion and slows the moving object. The slowing causes the Coriolis Effect to decrease. As shown by the thick arrow at Point *D*, the direction of airflow changes and air flows obliquely across isobars towards **[(*lower*)(*higher*)]** pressure.

17. It is the presence of the frictional force added to the pressure gradient force and the Coriolis Effect that causes air to spiral **[(*inward*)(*outward*)]** in surface-map Lows and outward in Highs.

An additional force acts on horizontally moving air if the isobars are curved. That force, called the *centripetal force*, is not treated in this investigation.

As directed by your course instructor, complete this investigation by either:

 1. *Going to the Current Weather Studies link on the course website, or*
 2. *Continuing to the Applications section for this investigation that immediately follows in this Investigations Manual.*

Investigation 8A: Applications

SURFACE WEATHER MAPS AND FORCES

Figure 1 is the surface weather map (Isobars, Fronts, Radar & Data) for 00Z 28 MAR 2010. At map time a vigorous spring storm system was crossing the south-central U.S. This was marked by two **L**s in southern Missouri. A second weaker storm was located over the Great Lakes area. Both storms were wedged between relatively fair-weather air masses, of which one was departing the eastern U.S. and the other entering the inter-mountain region of the West. As the southern storm traveled across the Southeastern U.S., thunderstorms produced prodigious rainfall amounts and some hail. Also, several tornadoes were spawned.

18. The wind at Wichita, Kansas, where the temperature was 43 °F and the dewpoint 36 °F, showed the air was moving generally <u>toward</u> the **[(*south*)(*west*)(*north*)(*east*)]** at about 25 knots.

19. Draw a short, straight line through Wichita that is perpendicular to the nearby 1008-mb isobar. Extend the line to the adjacent isobars (1012 and 1004). Place an arrow head on the end of the line toward lower pressure. Your arrow represents the direction of the **[(*Coriolis*)(*pressure gradient*)(*friction*)]** force acting on the air at Wichita.

20. This force acting on the wind flow at Wichita was directed generally toward the **[(*west-southwest*)(*west-northwest*)(*east-southeast*)(*north-northeast*)]**.

21. The wind direction at Wichita was **[(*parallel*)(*at an angle*)]** to the force arrow you drew perpendicular to the isobars.

22. Draw a small arrow at a right angle to the wind flow to represent the Coriolis effect acting on the wind flow at Wichita. It was directed generally toward the **[(*east-southeast*) (*west-southwest*)(*north-northwest*)(*east-northeast*)]**.

23. The force acting generally toward the north on the air, in the direction opposite to the air flow at Wichita, was the local **[(*Coriolis*)(*pressure gradient*)(*friction*)]** force.

24. Considering the Wichita conditions as part of the circulation about the south-central Low, the horizontal pressure gradient, Coriolis and friction forces combine to direct surface air flow around Northern Hemisphere low-pressure centers that is **[(*clockwise and outward*) (*counterclockwise and inward*)]**, consistent with the hand-twist model.

25. Now consider Grand Junction, in western Colorado. Grand Junction had a northwest wind of 10 knots and a temperature of 51 degrees. Grand Junction was influenced by the western High's circulation. Grand Junction's pressure gradient force (perpendicular to the local isobar) was directed generally <u>toward</u> the **[(*southeast*)(*northwest*)(*northeast*) (*southwest*)]**.

26. The Coriolis force at Grand Junction was directed generally toward the [(*southeast*) (*northwest*)(*northeast*)(*southwest*)].

27. The friction force at Grand Junction was toward the general direction of [(*southeast*) (*northwest*)(*northeast*)(*southwest*)].

28. Considering the Grand Junction conditions as part of the circulation about the western High, the horizontal pressure gradient, Coriolis and friction forces combine to direct surface air flow around the Northern Hemisphere high-pressure center that is [(*clockwise and outward*)(*counterclockwise and inward*)], consistent with the hand-twist model.

29. The horizontal pressure gradient force is the primary determiner of wind speed. Generally higher wind speeds are located where the pressure gradient is stronger. The spacing of isobars on the map implies the strength of the pressure gradient with closer spacing associated with stronger gradients. The stronger pressure gradients at this evening map time were located in the region centered over the [(*eastern Carolinas'*) (*northern Missouri-eastern Kansas*)] area.

The pressure patterns, including the pressure gradients they produce, are major determiners of local weather conditions including wind speeds and directions. However, local conditions can also influence winds, particularly when the broad-scale pressure patterns are not dominant. For example, light pressure gradients allow sea breeze winds to flow onshore from late mornings to early evenings while the flow can be offshore late evenings. It is probable that Boston has a sea breeze near the center of the eastern High's influence. And, wind may be channeled by local terrain independently of isobar orientations. Mountainous regions in the West with greater friction may hinder winds reaching a balance with large-scale pressure gradients.

Suggestions for further activities: The "Isobars, Fronts, Radar, & Data" map from the course website can be used to identify the directions of forces associated with winds in weather systems. An interesting challenge is to compare the weather maps such as we have used with those for locations in the Southern Hemisphere. For example, *http://www.bom.gov. au/weather/national/charts/synoptic.shtml* shows the latest surface analysis for Australia. Observations can be found at: *http://weather.noaa.gov/*, which may be plotted on the Australian map.

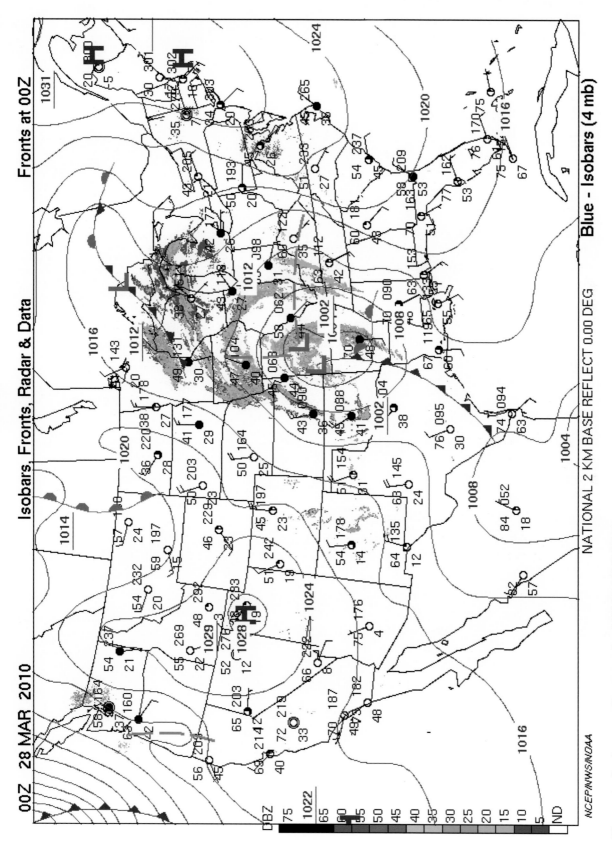

Figure 1. Surface weather map for 00Z 28 MAR 2010.

Investigation 8B:

UPPER-AIR WEATHER MAPS

Objectives:

Weather as reported on surface weather maps provides us primarily with a two-dimensional view of the state of the atmosphere, that is, weather conditions observed at Earth's surface. Atmospheric conditions reported on upper-air weather maps provide the third dimension, that is, conditions at various altitudes or pressure levels above the Earth's surface. For a more complete understanding of the weather, we need to consult both surface and upper-air weather maps.

After completing this investigation, you should be able to:

- Describe the topography of upper-air constant-pressure surfaces based on height contours, including the identification of Highs, Lows, ridges, and troughs.
- Identify the general relationship between height contours and the temperature of the underlying atmosphere.
- Describe the relationship between the height contours and wind direction on upper-air weather maps.

Introduction:

Upper-air weather maps differ from surface weather maps in several ways. Whereas surface weather conditions are plotted on a map of constant altitude (normally sea level) from observations that are collected at least hourly, upper-air weather conditions are commonly plotted on maps of constant air pressure from rawindsonde observations made twice a day. The altitude at which the particular pressure occurred is reported on these maps. An upper air observation is made by releasing a balloon-borne instrument package to the atmosphere. As the balloon rises, the air pressure decreases. The altitude at which the pressure of 850 mb occurs is referred to as the 850-mb height. Every 12 hours, upper-air maps are routinely drawn for various pressure levels including 850 mb, 700 mb, 500 mb, and 300 mb.

Plotted on upper-air maps are temperature (in °C), dewpoint (in °C), wind speed (in knots), wind direction, and height of the pressure surface above sea-level (coded in tens of meters). Become familiar with the upper-air station model depicted below. The upper air station is located at the end of the wind shaft opposite the speed "barbs" and/or "flags".

UPPER AIR STATION MODEL LEGEND
(500 mb)

Temperature (°C)   -28   548   Altitude (tens m) = 5480 m

Dewpoint (°C)   -42   Wind (southeast, 65 kts)

While the AMS course website provides upper air maps utilizing the station model above with the dewpoint shown directly, it should be noted that several NOAA and other web maps display the dewpoint depression instead. The *dewpoint depression* is the number of Celsius degrees that the dewpoint is <u>below</u> the plotted temperature. To prevent confusion as to which (dewpoint or dewpoint depression) is being reported when interpreting upper air maps from different sources, keep in mind that the dewpoint can never be greater than the air temperature, so the dewpoint depression must <u>always</u> be a positive number or 0.

Whether the dewpoint or dewpoint depression is displayed, meteorologists generally assume that clouds are present (at the station or within the region) when the dewpoint is within 5 Celsius degrees of the air temperature, *i.e.* the dewpoint depression is 5 or less.

The plotted altitude of the pressure level of the map is a coded value. That is, only the three most significant digits of the value are plotted. To decode the plotted value the following table shows the Standard Atmosphere altitude of the pressure surface, and the missing digits needed to decode the three plotted numbers (xxx).

Upper Air Map (mb)	Standard Altitude (m)	Coded Digits
850	1457	1xxx
700	3012	(2 or 3)xxx
500	5579	xxx0
300	9164	xxx0

The 700-mb level may lie below 3000 m necessitating placing a 2 in front of the plotted values to make the meaningful choice in the decode.

1. **Figure 1** is the 500-mb map for 00Z 12 APR 2010. Meteorologists frequently refer to 500-mb maps because winds at that level generally steer weather systems across Earth's surface. Hence, the so-called *steering winds* at 500 mb can be used to predict the track of a low-pressure system.

   Solid lines on the 500-mb map join locations where the 500-mb pressure level is at the same altitude. These lines, called *contours of height*, are drawn at intervals of 60 m. The coded height values on the map are in **tens** of meters. Contour values are in **whole** meters. On the Figure 1 map, the lowest height reported at an individual station for a pressure reading of 500-mb was [(*5220*)(*5350*)(*5380*)(*5420*)] m.

2. The highest reported 500-mb height at an individual station on the Figure 1 map was [(*5730*)(*5750*)(*5800*)(*5830*)] m.

3. The 500-mb map and other constant-pressure upper-air maps are actually topographic maps that give form or shape to an imaginary surface on which the air pressure is everywhere the same. That is, the contour pattern reveals the "hills" and "valleys" of the constant-pressure surface. The contour pattern of the Figure 1 map indicates that, in general, the 500-mb surface (the surface where the air pressure is everywhere 500 mb) is at a [(*higher*)(*lower*)] altitude in southern Canada than in the southern U.S.

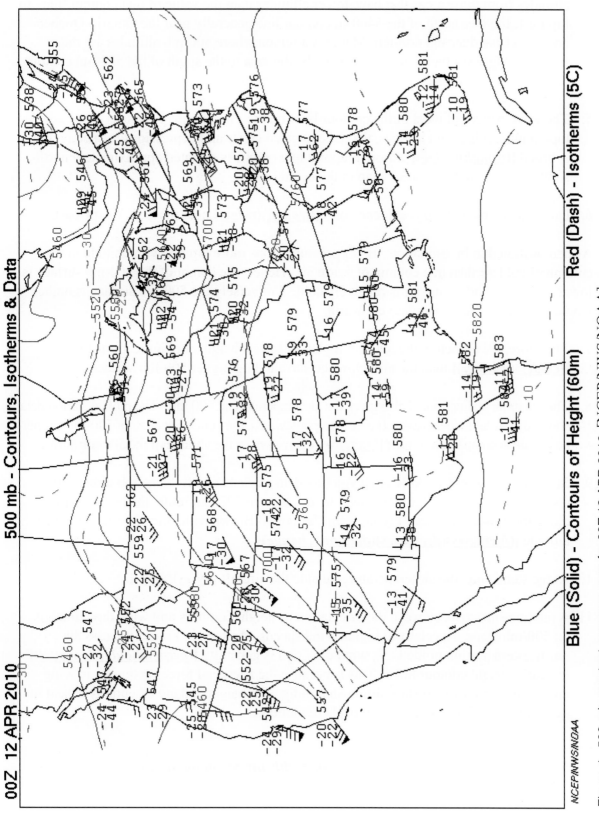

Figure 1. 500-mb constant-pressure map for 00Z 12 APR 2010. [NCEP/NWS/NOAA].

4. Contour lines on constant-pressure upper-air maps separate regions that have higher altitudes from those areas that have lower altitudes than the value of that contour line. In Figure 1, the area north of the 5460-m contour line generally near the Ontario-Quebec border in Canadian and northern Maine is a region where 500-mb altitudes are the [(*lowest*)(*highest*)] on the map. Conversely, the area to the south of the 5820-m contour across southern Texas is the highest on the same map.

5. The wave pattern of most of the contour lines on the Figure 1 map consists of topographic ridges and troughs, that is, elongated crests and depressions, respectively. A broad [(*trough*)(*ridge*)] whose midpoint line or axis is oriented approximately north-south appears on the Figure 1 map in the central U.S.

6. On the same map, there is evidence of a [(*trough*)(*ridge*)] near the U.S. West Coast.

As demonstrated in Investigation 5B, air pressure drops more rapidly with altitude in a column of cold air than in a column of warm air. Hence, the height of the 500-mb surface is lower where the underlying air is relatively cold. Conversely the 500-mb surface is higher where the underlying air is relatively warm.

7. Therefore, the air below the 500-mb region of lowest heights in Figure 1 must be [(*colder*)(*warmer*)] than the air below the surrounding higher 500-mb surfaces.

8. The upper air station model also gives the air temperature at 500 mb. The plotted station data show that, as latitude increases (*i.e.* moving poleward), the general decline of 500-mb temperatures are accompanied by a(n) [(*increase*)(*decrease*)] in the altitude of the 500-mb surface.

9. Suppose that at 00Z 12 APR 2010 you board an airplane and fly non-stop directly from Caribou, in northern ME to Brownsville, in southern TX. En route, the plane cruises along the 500-mb surface. Flying from Caribou to Brownsville, the aircraft's cruising altitude [(*increases*)(*decreases*)(*does not change*)].

10. At the same time, the air temperature outside the aircraft [(*rises*)(*falls*)].

11. A relationship exists between the orientation of height contours and wind direction on 500-mb maps, especially at higher wind speeds. As seen in Figure 1 across the northeastern portion of the U.S., wind direction is generally [(*perpendicular*)(*parallel*)] to nearby height contour lines. This is because the frictional forces acting on moving air at and near Earth's surface diminish rapidly with height and are essentially absent in determining middle and upper atmosphere motions.

As directed by your course instructor, complete this investigation by either:

1. *Going to the Current Weather Studies link on the course website, or*
2. *Continuing to the Applications section for this investigation that immediately follows in this Investigations Manual.*

Investigation 8B: Applications

UPPER-AIR WEATHER MAPS

Figure 2 is the 500-mb constant-pressure map for 00Z 28 MAR 2010. These were the upper-air conditions over the coterminous U.S. and adjacent areas of Canada and Mexico <u>at the same time</u> as the conditions shown on the Figure 1 surface map of *Investigation 8A*. The 500-mb conditions are those in the middle troposphere associated with the surface conditions, including the storm systems and air masses that dominate the surface map. Weather systems extend well into the troposphere and a three-dimensional understanding of them is necessary for accurate predictions.

Maps of upper-atmospheric conditions are made twice each day at 00Z and 12Z, and are based on rawinsonde reports. Those displayed on the course website are from the radiosondes launched from about 70 stations across the continental U.S. and Canada/ Mexico areas. On an upper-air map, the temperature, dewpoint, height and wind data from a station's rawinsonde report at that pressure are plotted around each station location (at the forward end of the wind arrow) in an upper-air station model format, as discussed earlier in this Investigation and the *User's Guide*, linked from the **Extras** section of the course website.

12. On the Figure 2, 500-mb map, the plotted report for Springfield, in southern Missouri, shows that at 500 mb over the station, the temperature was **[(_–18_)(_–22_)(_–30_)]** °C.

13. The dewpoint at 500 mb over Springfield was **[(_–23_)(_–36_)(_–50_)]** °C.

14. Recalling that the heights plotted at individual stations on 500-mb maps are in <u>tens</u> of meters (add a **0** to the three plotted digits), the height at which 500 mb occurred over Springfield was **[(_5710_)(_5640_)(_5490_)]** meters.

15. The wind at Springfield was from the southwest at about **[(_20_)(_35_)(_65_)]** knots. [*Note: When winds are 50 knots or greater, a pennant (triangular flag) is placed on the station's wind shaft for a 50-kt increment along with the usual long and short barbs.*]

16. The following data were from a radiosonde report at another station's 500-mb level at 00Z 28 MAR 2010:
    Height (m): 5740, temperature (°C): –17, dewpoint (°C): –62,
    wind direction (deg. from N): 255 (*i.e.* WSW), wind speed (kts): 25.

    Examining the stations plotted on the 500-mb map in Image 1 shows this station to be **[(_Charleston, SC_)(_Grand Island, NE_)]**. [Current radiosonde reports containing upper air data can be found from the **Upper Air** section, "Upper Air Data - Text" on the course website.]

The pattern of 500-mb heights (heights above sea level where the air pressure is 500 mb as found by the radiosondes at that time) can be shown by contour lines. To better visualize the contour pattern plotted by the computer on the Figure 2 map, highlight the 5580-m contour by tracing over it. [The 5580-m contour curves from the northwest tip of Washington State, across western Canada, southward to northern Texas, then northeastward to cross New York State and finally exit the U.S. over Massachusetts.]

17. The contour pattern of the 500-mb map has [(*a deep trough in the central U.S.*)(*a ridge over the western U.S.*)(*a ridge along the Atlantic Coast and eastern U.S.*)(*all of these listed ridge and trough features*)].

18. The Figure 2, 500-mb map also shows that, where contour lines are relatively close, such as along the western Gulf Coast states, wind speeds are relatively [(*low*)(*high*)]. This principle corresponds to that of the spacing of isobars and wind speeds on surface maps.

19. At upper levels, the wind directions are also related to the contours. That is, where winds are relatively fast, the winds are generally [(*"parallel" to the contours*)(*directed across the contours at large angles*)]. The absence of friction at upper levels means that flow is controlled mainly by the pressure gradient and Coriolis forces. Therefore winds are generally along the contours on upper level maps as opposed to the inward circulations with Lows and outward with Highs seen on surface maps.

20. Using the station values as well as the contour pattern, compare the heights at Springfield, MO to those of San Francisco, CA in the west and Cape Hatteras, NC in the east. At these comparable latitudes, the 500-mb heights were relatively [(*lower*)(*higher*)] over Springfield area compared to those over San Francisco and Cape Hatteras.

21. Also compare the temperatures for these stations. The 500-mb temperatures are relatively [(*lower*)(*higher*)] over Springfield compared to San Francisco and Cape Hatteras. As you recall, the relation of column temperatures to heights of pressure surfaces was examined using conceptual "pressure blocks" in Investigation 5B.

22. The direction of movement of surface low-pressure centers can often be anticipated from the 500-mb wind flow directions over their locations. From the Figure 1 surface weather map of *Investigation 8A* as reference, place the low-pressure center of the storm at that time with a bold *L* in southern Missouri on Figure 2. Using the general wind directions at 500 mb over southern Missouri to northern Louisiana as guides, one could predict the movement of the surface Low in the next day or so as being toward the [(*southeast*)(*northeast*)(*northwest*)(*southwest*)].

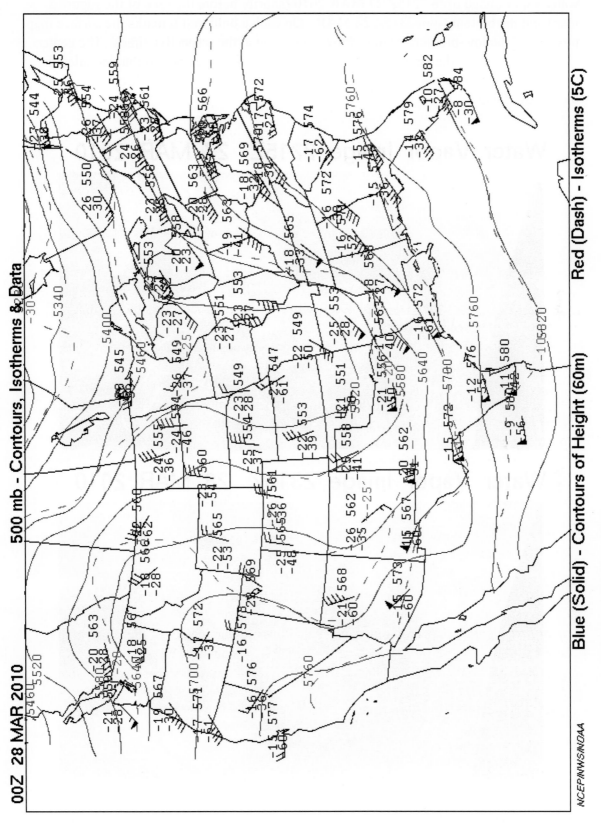

Figure 2. 500 mb constant pressure map for 00Z 28 MAR 2010.

23. **Figure 3** is a composite image of two water vapor satellite views. The top view is the water vapor image from 2215Z 27 MAR 2010 (shortly before the time of the Figure 2 map) and the bottom from 2315Z 28 MAR. On each a bold, red L marks the surface map position of the low-pressure center of the circulation at the respective times. The center of the circulation of the storm system over the twenty-five hour period that included the Figure 2 500-mb map [(___did___)(___did not___)] generally follow your predicted direction based on the Figure 2 winds.

Water Vapor Image 2215Z 27 MAR 2010

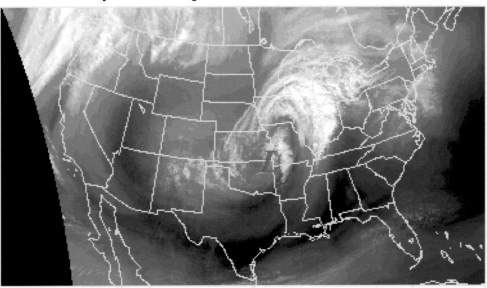

Water Vapor Image 2315Z 28 MAR 2010

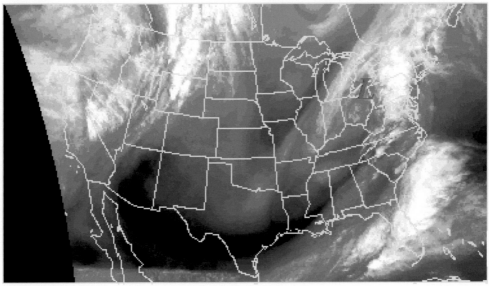

Figure 3.
Composite of water vapor satellite images at 2215Z 27 MAR 2010 (top) and 2315Z 28 MAR 2010 (bottom).

Figure 4 is the surface weather map (Isobars, Fronts, Radar & Data) for 00Z 31 MAR 2010, Tuesday evening. The storm system that was centered in the south-central U.S. three days earlier on Saturday evening had moved northeastward. The low-pressure center was located over Long Island, NY. This low-pressure center had become relatively stationary over the preceding twenty-four hours. The precipitation shield of the storm, as shown by the radar shadings, was also relatively stationary bring copious rainfalls, especially across New England leading to some severe local flooding.

Also shown stretched across the western U.S. is a cold front extending from a *triple point* of occluded, warm and cold fronts in southern Canada just north of the North Dakota-Minnesota border to southern California. The strong winds directed northward across the central Plains are a result of the strong horizontal pressure gradient over the central U.S. evidenced by the closeness of isobars. Note that these early evening spring temperatures reached the seventies all the way to Minnesota showing the strong warm air advection.

24. Compare the surface temperatures at Albany, in eastern NY, Minneapolis, in southeastern MN, and Portland, OR. The central U.S. was generally **[(*colder*)(*warmer*)]** than the two coasts.

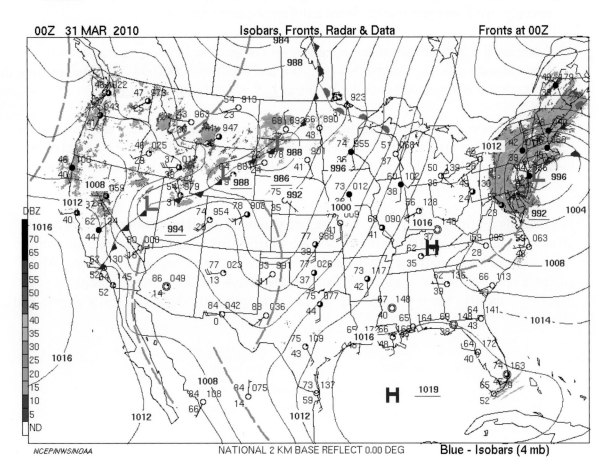

Figure 4.
Surface weather map for 00Z 31 MAR 2010.

Figure 5 is the 500-mb constant-pressure map for 00Z 31 MAR 2010. These were the middle tropospheric conditions that correspond to the surface weather of the Figure 4 map. They are also the upper-air conditions seventy-two hours following the Figure 2, 500-mb map.

25. Visualize the 500-mb contour pattern as you did before by highlighting the 5640-m contour in Figure 5 from south-central California to southern Wisconsin and then to South Carolina. The general 500-mb contour pattern of Image 3 is one of [(***East Coast trough and central U.S. ridge***) (***central U.S. trough and East Coast ridge***)]. There is also evidence that a trough is beginning to be displayed along the West Coast.

26. The closed 5460-m contour off the New Jersey coast [(***does***)(***does not***)] generally coincide with the location of the surface Low surrounded by the 996-mb isobar on the Figure 4 map. The "stacked" position of the low features at both levels is indicative of a stationary or slowly moving system, in this case one that brought considerable moisture over New England leading to the flooding. Note the onshore flow of winds at both levels over the Massachusetts to Maine region providing Atlantic moisture. Also, the stacking means upper-level winds are not moving the systems rapidly eastward as was the case seventy-two hours previously. What a difference three days makes!

27. Compare the 500-mb station temperatures for comparable latitudes at Brookhaven on Long Island (–20 °C), Omaha, NE (–12 °C) and Medford, OR (–34 °C). The warmest air at 500 mb was associated with the [(***central U.S. ridge***)(***eastern trough***)].

Middle and upper tropospheric conditions are inextricably linked with the surface weather features. They are involved in the development and movement of weather systems over the Earth. We will consider these relationships along with upper tropospheric maps and conditions in Investigation 9A.

Suggestions for further activities: You might try making a height-contour analysis by printing an unanalyzed 500-mb map ("500 mb - Data") from the website. You can then compare your hand-analyzed pattern to the computer-analyzed map with contours. Also, compare upper-air map patterns to surface weather maps and the weather conditions you experience locally. See if you can link upper-air troughs and ridges with surface Lows and Highs.

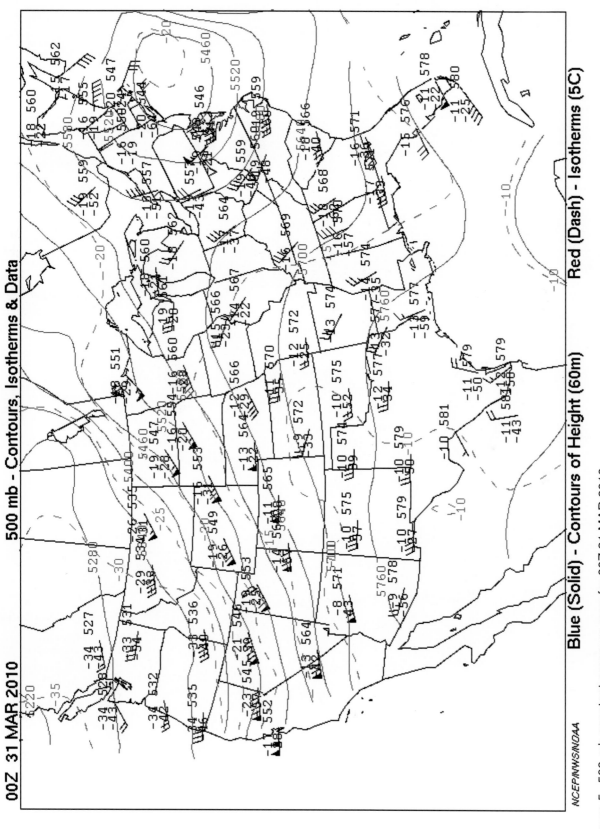

Figure 5. 500-mb constant-pressure map for 00Z 31 MAR 2010.

Investigation 9A:

WESTERLIES AND THE JET STREAM

Objectives:

At the planetary (global) scale in the middle and upper troposphere, the prevailing upper-air westerlies encircle middle latitudes in a wave-like pattern. These winds are important components of day-to-day weather in that they steer storm systems from one place to another and are ultimately responsible for the movement of air masses. Surveying the basic characteristics of these upper-air tropospheric westerlies is key to understanding the variability of midlatitude weather.

Relatively narrow "rivers" of strong winds, called *jet streams*, exist at middle and upper tropospheric levels at different times within the westerlies. Jet streams that occur over the polar front and near the tropopause have important influence on the weather of middle latitudes. These so-called polar-front jet streams exist where relatively cold air at higher latitudes comes in contact with warm air from lower latitudes. In addition, these jet streams provide upper-air support for the development of low pressure systems.

After completing this investigation, you should be able to:

- Describe the wave patterns exhibited by the meandering upper-air westerlies.
- Determine the location of the polar-front jet stream on an upper-air weather map.
- Explain the general relationships between the jet stream in the upper-air westerlies and the paths air masses and storms take.
- Describe how atmospheric temperature patterns are associated with the upper-air circulation and the jet stream.

Introduction:

1. The upper-air westerlies flow generally from west-to-east around the planet in a wave-like pattern of ridges and troughs as shown below. Ridges are topographic crests and troughs are elongated depressions on constant-pressure surfaces. (Refer to Investigation 8B to review features on upper-air maps, including ridges and troughs.) In the diagram below, "H" locates a ridge and "L" locates troughs.

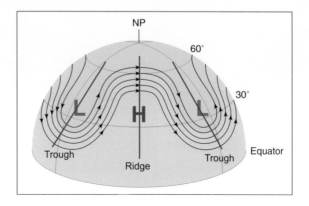

The Northern Hemisphere's upper-air westerlies exhibit clockwise (anticyclonic) curvature in ridges. As shown in the drawing above, a line can be drawn that divides a ridge into two roughly symmetrical sectors. The line is known as a *ridge line*. Note that west of the ridge line, winds are from the southwest (a warm weather direction) and east of the ridge line, winds are from the northwest (a cold weather direction). We can conclude that winds to the west of a ridge line favor [(*cold*)(*warm*)] air advection while winds to the east of a ridge line favor cold air advection.

2. The upper-air westerlies curve counterclockwise (cyclonic) in troughs. As shown in the drawing above, a line can be drawn that divides a trough into two roughly symmetrical sectors. The line is known as a *trough line*. Note that west of the trough line, winds are from the northwest (a cold weather direction) and east of the trough line, winds are from the southwest (a warm weather direction). We conclude that winds to the west of a trough line favor [(*cold*)(*warm*)] air advection while winds to the east of a trough line favor warm air advection.

    Ridges and troughs usually progress from west to east over time so that as a ridge line shifts eastward, a location that had been experiencing cold air advection then experiences warm air advection, and a location that had been experiencing warm air advection then experiences cold air advection.

3. Upper-air winds steer low-pressure systems as well as air masses. A surface Low that is centered to the east of a trough line and west of a ridge line will be expected to move toward the [(*northeast*)(*southwest*)].

The wavy pattern of the upper-air westerlies consists of ridges alternating with troughs. The distance between successive ridge lines or, equivalently, between successive trough lines is the *wavelength*. At any one time, usually between 2 and 5 waves encircle the Earth in the middle latitudes.

With time, the wave pattern of the upper-air westerlies changes. These changes may involve a change in the number of waves, the wavelength, or the amplitude of the wave. At one extreme, shown in the left drawing below, upper-air westerlies blow almost directly from west to east with little sign of ridges or troughs. This westerly flow pattern is described as *zonal*. At the other extreme, shown in the right drawing below, upper-air westerlies blow in huge north/

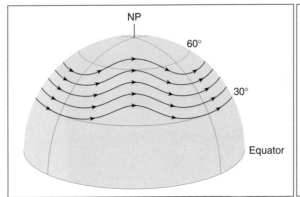
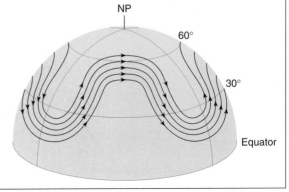

south loops with high amplitude ridges and troughs. This westerly flow pattern is described as *meridional*. The circulation patterns displayed below are opposite extremes of many possible patterns commonly exhibited by middle latitude upper-air westerly waves.

4. When the upper-air westerly flow pattern is zonal, the source region for much of the air over the coterminous U.S. is the Pacific Ocean. On the other hand, when the upper-air westerly flow pattern is meridional, the source regions for air masses over the lower 48 states are Canada (where winds are from the northwest) or Mexico or the Gulf of Mexico (where winds are from the southwest). Hence, from west to east across the lower 48 states, temperatures are likely to be more variable with a **[(*zonal*)(*meridional*)]** flow pattern.

5. Fundamental to the formation of the polar-front jet stream within the westerlies is the physical property that warm air is less dense than cold air when both are at the same pressure. Air pressure drops **[(*more*)(*less*)]** rapidly with increasing altitude in cold air than in warm air.

The polar front marks the lower-atmosphere boundary between higher latitude cold air and lower latitude warm air. This temperature contrast extends from Earth's surface up to the altitude of the polar-front jet stream. As demonstrated in Investigation 5B, the effect of temperature on air density means that the air pressure at any given altitude above the surface is higher in the warm air column than in the cold air column. Hence, a horizontal pressure gradient is directed across the front from the warm side toward the cold side. In response, the horizontal wind initially blows from warm air toward cold air but is soon deflected to the right by the Coriolis Effect. Consequently, the wind blows parallel to the polar front with the cold air to the left when facing in the direction towards which the air is flowing (in the Northern Hemisphere). Furthermore, where cold and warm air reside side by side, the magnitude of the horizontal pressure gradient increases with increasing altitude. This causes the horizontal wind to strengthen with altitude and reach its maximum speed in the polar-front jet stream.

6. In the Northern Hemisphere, when the polar-front jet stream is south of a locality, the weather at that location is relatively **[(*warm*)(*cold*)]**.

7. As a component of the planetary-scale upper-air westerlies and similar to the winds at 500 mb, the polar-front jet stream steers low pressure systems. Hence, middle latitude storms generally move from **[(*west to east*)(*east to west*)]**.

Examine **Figure 1**, the upper-air map of winds at the 300-mb level at 12Z 05 MAY 2010 (average 300-mb height is about 9 km above sea-level). **Using a pencil, lightly shade stations that have wind speeds of 70 knots or higher (triangular pennant and two long barbs) across the northern half of the U.S. Then lightly shade the area between such adjoining stations to form a broad band where wind speeds are 70 knots or higher. Draw a dark, heavy, smooth, curved arrow across the map through the middle of the broad band. Add an arrowhead to represent wind direction.** The large arrow you drew on your map approximates the location of the polar-front jet stream across the coterminous 48 states.

Examine **Figure 2**, the upper-air map of winds at the 300-mb level at 12Z on 02 MAR 2010, upper tropospheric conditions about two months earlier than the Figure 1 map. **Using a pencil, again lightly shade the region(s) where winds are at least 70 knots. Finally, draw a heavy dark arrow and arrowhead through the high-speed core of the jet stream winds.** In Figure 2, regions of the high speed winds are located in several unconnected patches.

8. The pattern of winds in Figure 2 indicates a [(*ridge*)(*trough*)] over the east-central states. The flow pattern in the same region is meridional.

9. On average, one would expect temperatures at similar latitudes to be similar. Zonal wind patterns, as in Figure 1, exhibit this relationship. However, given the wind pattern of Figure 2, at 12Z on 02 MAR 2010, surface air temperatures over western Idaho are likely to be [(*lower*)(*higher*)] than surface temperatures over Wisconsin. This is reflected in the 300-mb temperatures plotted on the map.

10. Across the coterminous U.S., Lows tend to follow the path of the polar front jet stream. At Figure 2 map time, a storm was centered over the Florida panhandle. From Figure 2, this storm was likely to move towards [(*Louisiana*)(*Illinois*)(*New Jersey*)].

11. Knowledge of the location of the jet stream and upper-air winds in general is very important for commercial aviation and can result in fuel savings and shorter flight times. At Figure 2 map time, an airline flight from Atlanta, Georgia to Boston, Massachusetts would take [(*less*)(*more*)] time than a flight along the same route from Boston to Atlanta.

As directed by your course instructor, complete this investigation by either:

1. *Going to the Current Weather Studies link on the course website, or*
2. *Continuing to the Applications section for this investigation that immediately follows in this Investigations Manual.*

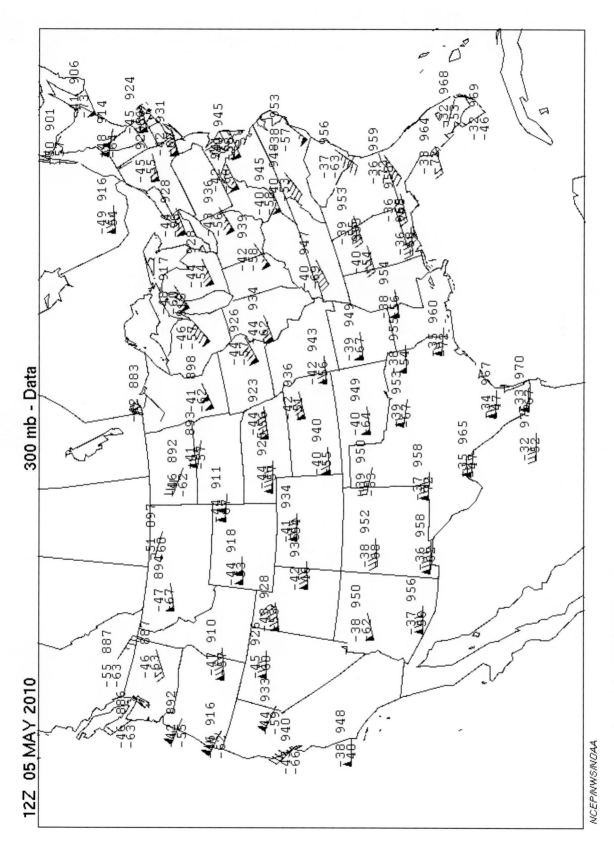

Figure 1. 300-mb winds map for 12Z 05 MAY 2010.

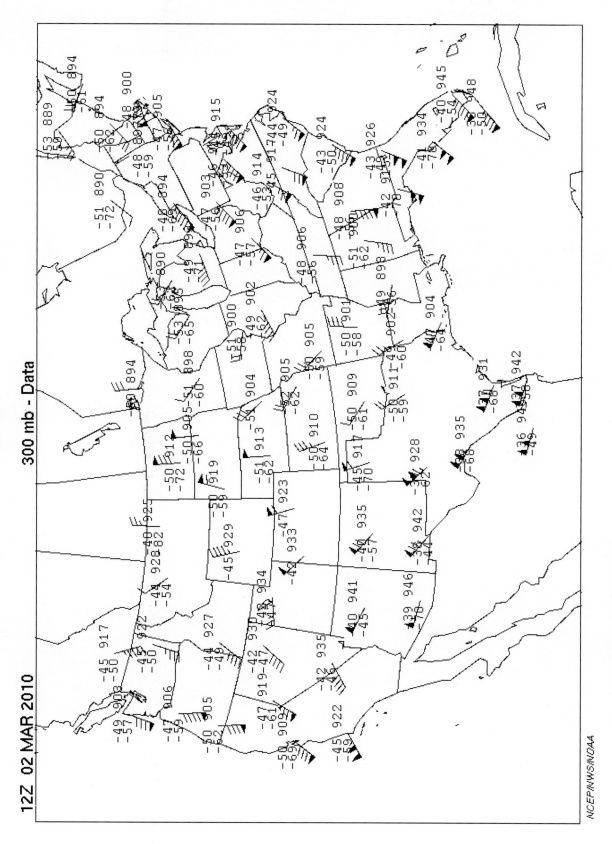

Figure 2. 300-mb winds map for 12Z on 02 MAR 2010.

Investigation 9A: Applications

WESTERLIES AND THE JET STREAM

In early April 2010, a storm system producing heavy rains moved northeastward from the central U.S. While the circulation center of the storm traveled into Canada, the trailing cold front swept across the eastern U.S.

Figure 3 is the surface weather map for 00Z 03 APR 2010. At that time, the storm's intense low-pressure center was located over International Falls, MN on the U.S.-Canadian border with a long trailing cold front over the central U.S. to northern Mexico. Also, a warm front extended from the low center eastward across Canada's Ontario Province. A short, secondary cold front was shown trailing the first, from western Iowa to the Texas panhandle. The second front marked the leading edge of even cooler and drier air. A broad band of intense precipitation stretched between Wisconsin and northeastern Texas in the warm sector of the storm. Another system had come ashore along the British Columbia to California coasts marked by the low-pressure center over Vancouver Island and an occluded front arching southeastward to California where it was shown as a cold front across the state to the Pacific.

12. Based on station models from North Dakota to Wisconsin, the winds indicated that the flow about the central U.S. Low center was [(***clockwise and outward***)(***counterclockwise and inward***)]. This was also the case evidenced by the station models in Washington State about the western Low.

13. Station models along the East Coast showed that the broad circulation about the western side of an elongated High whose center was shown located over Cape Cod, Massachusetts, was [(***clockwise and outward***)(***counterclockwise and inward***)]. This was also the case generally evidenced by the station models about another high-pressure area displayed over the Four Corners of UT-CO-NM-AZ.

14. From the number of isobars plotted about the International Falls and Vancouver Lows compared to those about the Cape Cod and Four Corners Highs, one can conclude that relatively [(***stronger***)(***weaker***)] pressure gradients are associated with Lows than with Highs. In general, wind speeds are somewhat faster near the Low centers as compared to those about the broad central areas of Highs.

15. **Figure 4** is the 500-mb constant-pressure map for Friday evening 00Z 03 APR 2010, the same time as the Figure 3 surface map. The 500-mb pressure level is found in the [(***lower***)(***middle***)(***upper***)] troposphere.

16. The dominant upper-air flow pattern feature at 00Z 03 APR 2010 as shown on the 500-mb map was one of a [(***trough***)(***ridge***)] in the west-central U.S. Additionally, there existed a strong ridge along the East Coast and a weaker ridge over the Rocky Mountain region of the West.

17. Recall that the heights at which the radiosondes detected 500-mb of pressure are reported in the upper right position of the upper-air station models on the map. Also, the heights are plotted in <u>tens</u> of meters, so that a "*0*" needs to be added to the digits for the actual height. The height of the 500-mb pressure level plotted on the Bismarck, North Dakota station model was **[(*5290*)(*5320*)(*5480*)]** m.

Taking the southern-most extension of each contour on the 500-mb map from east-central South Dakota to west Texas, draw a dashed, curved line through those southern extensions. This dashed line marks the *trough line* where winds generally change from being northwesterly to the west of that line to being southwesterly east of the line. Also, sketch the general position of the extensive central-U.S. cold front from the surface map onto this 500-mb map.

18. The cold front extending from the International Falls Low was generally located to the **[(*east*)(*west*)]** of the trough line on the 500-mb map. The surface Low center shown on Figure 3 is also located to the east of the closed 5340-m contour plotted over the Dakotas and extreme southern Canada of the Figure 4 map.

19. The highest wind speed plotted on the 500-mb map is at Midland, TX where there are two pennants for a total speed of **[(*75*)(*100*)(*115*)]** kts generally from the northwest, much higher than surface wind speeds.

20. **Figure 5** is the 300-mb constant-pressure map for 00Z 03 APR 2010, the same time as the surface and 500-mb maps. The general 300-mb contour pattern at 00Z on 3 APR across the west-central portion of the U.S. displayed a **[(*trough*)(*ridge*)]**. The western and eastern ridges that were evidenced on the 500-mb map were also present.

21. The heights of the undulating 300-mb pressure surface were within several hundred meters of **[(*5500*)(*9200*)(*12,500*)]** meters. The 300-mb level occurs in the <u>upper</u> troposphere.

22. Comparing the heights of the 300-mb pressure surface from south to north on the map, as latitude increases (one moves poleward) the height of the 300-mb surface generally **[(*increases*)(*remains the same*)(*decreases*)]**. There is a corresponding change in temperature, confirming that pressure decreases more rapidly in the vertical in colder columns of air than in warmer columns (again recall Investigation 5B).

23. The dashed (brown on-screen) lines on the 300-mb map are lines of equal wind speed, *isotachs*, drawn at 20-knot intervals for wind speeds of 30 knots and higher. The highest wind speed at 300 mb plotted on the Image 3 map is at Oklahoma City, OK. The two pennants, three long barbs and a short barb indicate a speed of about **[(*75*)(*135*)(*195*)]** knots.

24. Using the conventional wind speed threshold of 70 knots for defining the existence of a jet stream, there **[(*was*)(*was not*)]** evidence of a jet stream indicated on the 00Z 03 APR 2010 300-mb constant-pressure map stretching across the U.S.

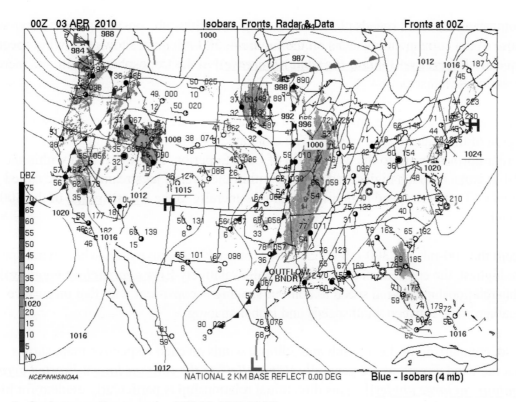

Figure 3. Surface weather map for 00Z 03 APR 2010.

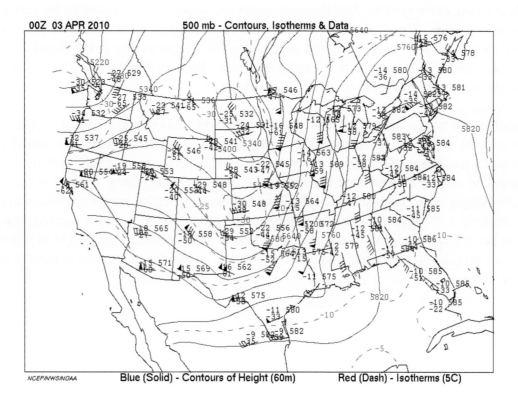

Figure 4. 500-mb constant-pressure map for 00Z 03 APR 2010.

25. Comparing the wind speeds plotted on the Figure 5, 300-mb constant-pressure map with those of the corresponding areas on the Figure 4, 500-mb map, shows that wind speeds typically [(***decrease***)(***remain the same***)(***increase***)] in the troposphere as altitude increases (pressure decreases).

26. There is a 110-kt isotach on the 300-mb constant-pressure map that encompasses the region of the highest wind speeds. The elongated oval band is from the Oklahoma-Texas border northeastward to the western tip of Lake Superior. Shade that area within the 110-kt isotach. Such high speed regions within the overall jet stream flow are called *jet streaks*. That high speed region was located generally to the [(***west***)(***east***)] of the 300-mb trough axis.

27. From the 500-mb and 300-mb winds shown on each respective map, where wind speeds are highest, the contour spacings are generally relatively [(***far apart***)(***close together***)]. This relationship of wind speed to contour spacing is consistent with that on surface weather maps between wind speeds and isobar spacing.

28. Also, comparison of the 500-mb and 300-mb winds on each respective map, shows that the wind directions are generally [(***"parallel" to the contour lines***)(***directed across contours at large angles***)]. This directional relationship is particularly evident for higher wind speeds.

While not examined in detail, these maps also show that there is a trough associated with the West Coast storm system. And there are upper-air ridges corresponding to the surface high-pressure areas. Upper air flow changes from zonal to meridional, and back again, with passing storm systems. And jet streams accompany the storms as they develop and dissipate. They also wander from south to north and back again. You can watch these changes by following the upper level maps on the course website.

Suggestions for further activities: You might try shading areas of highest wind speeds on 300-mb charts to identify jet streaks. *Jet streaks* are regions of accelerated wind speed along the axis of a jet stream. See if you can spot relationships between jet streaks and the location of associated surface low-pressure systems. Also, for developing storm systems in the central U.S., you might see if the positions of the low pressure/low height centers are successively more westerly with height (surface to 700 mb to 500 mb to 300 mb), as is expected with cold air being advected southward on the west side of storm centers. You might fit this with the earlier investigation of warm and cold air columns and their relationship to heights of pressure levels.

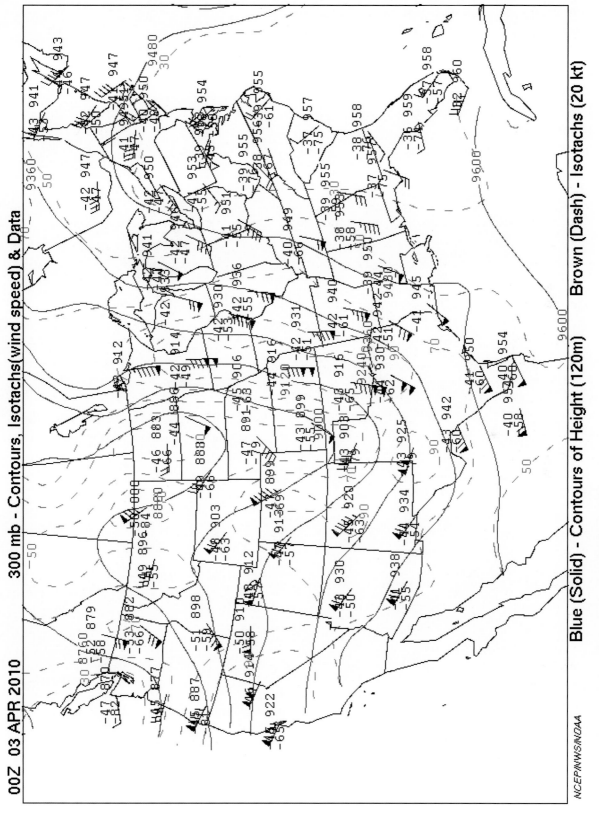

Figure 5. 300-mb constant-pressure map for 00Z 03 APR 2010.

¡EL NIÑO!

Objectives:

The tropical Pacific Ocean covers nearly one-fifth of Earth's surface area and stretches about one-third of the way around the globe. This vast expanse of ocean and the overlying atmosphere form a coupled system that makes its presence felt far beyond its boundaries. Its influence on world-wide weather and climate can have major ecological, societal, and economic consequences. Every 3 to 7 years this coupled ocean/atmosphere system takes on conditions termed El Niño, which typically persists for 12 to 18 months and may alternate with the less frequent La Niña. El Niño is one example of the variability of weather and climate on time and spatial scales that go beyond the basic weather map.

After completing this investigation, you should be able to:

- Describe the neutral (long-term average) conditions of the tropical Pacific Ocean and atmosphere.
- Compare El Niño and La Niña conditions to neutral conditions.
- Explain how atmospheric conditions during El Niño are transmitted beyond the tropical Pacific area.

Introduction:

Tropical Pacific during Neutral (Long-Term Average) Conditions

1. Examine **Figure 1**, the neutral (long-term average or normal) conditions in the tropical Pacific Ocean from about Borneo in the western Pacific Ocean to the west coast of South America (greatly exaggerated in the vertical scale). The scene depicts the ocean surface with atmosphere above and a cross-section of the ocean below. Fair weather appears in the eastern tropical Pacific (near 80 degrees W) while the cloud diagram implies that [(**_fair_**)(**_stormy_**)] weather prevails in the western Pacific (near 120 degrees E).

2. The large-scale motions in the atmosphere show a convection cell (convective loop). The bold dark arrows show that air is rising in the stormy weather area of the western Pacific and [(**_rising_**)(**_sinking_**)] in the eastern tropical Pacific.

3. The bold black arrow along the ocean surface in the convective loop represents the _trade winds_ and points in the direction toward which the prevailing winds are blowing in the equatorial region. As indicated by the arrows, winds during neutral (long-term average) conditions blow <u>toward</u> the [(**_east_**)(**_west_**)] along the equator.

4. The large white, open arrows provide surface ocean current information. The surface current arrows indicate that during neutral conditions, surface water flows towards the [(**_east_**)(**_west_**)] driven by the prevailing winds.

5. Colored areas on the top of the block diagram portion of the figure denote sea surface temperatures (SST) during neutral conditions. The red colored area in the western Pacific denotes the highest SST. These highest SST occur under [(*considerable cloudiness*)(*clear skies*)] in the tropical Pacific. This SST pattern is caused by relatively strong trade winds pushing sun-warmed surface water westward, as indicated by the direction of surface currents.

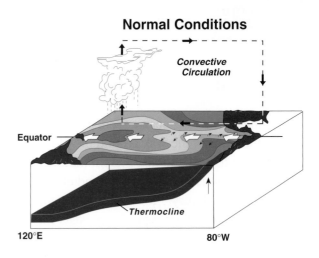

Figure 1.
Atmospheric-oceanic block diagram of Neutral ("Normal") Conditions in the tropical Pacific Ocean.

6. Strong trade winds also cause the warm surface waters to pile up in the western tropical Pacific so that the sea surface in the western Pacific is somewhat higher than in the eastern Pacific. Transport of surface waters to the west also causes the *thermocline* (the transition zone between warm surface water and cold deep water shown by the blue layer in the ocean side view) to be [(*deeper*) (*shallower*)] in the eastern tropical Pacific than in the western Pacific.

7. Warm surface water transported by the wind away from the South American coast is replaced by cold water rising from below in a process called *upwelling*. Upwelling of cold deep water results in relatively [(*high*)(*low*)] SST in the eastern Pacific compared to the western Pacific.

8. Cold surface water cools the air above it, which leads to increases in the surface air pressure. Warm surface water adds heat and water vapor to the atmosphere, lowering the surface air pressure. These air-sea interactions result in tropical surface air pressure being highest in the [(*eastern*)(*western*)] tropical Pacific.

9. Whenever air pressure changes over distance, a force will move air from where the pressure is relatively high to where pressure is relatively low. The trade winds blow from east to the west because from east to west the surface air pressure [(*increases*)(*decreases*)].

10. Rainfall in the tropical Pacific is also related to SST patterns. There are reasons for this relationship. The higher the SST, the greater the rate of evaporation of seawater and the more vigorous the atmospheric convection. Consequently, during neutral conditions, rainfall is greatest in the western tropical Pacific where SST are [(*highest*)(*lowest*)].

Tropical Pacific During *El Niño*

11. **Figure 2** shows atmospheric and oceanic conditions during *El Niño*. Compared to Figure 1 (neutral or long-term average conditions), the area of stormy weather during El Niño has moved [(*eastward*)(*westward*)]. While no two El Niño episodes are exactly alike,

all of them exhibit most of the characteristics shown in the El Niño schematic of Figure 2. With the onset of El Niño, tropical surface air pressure patterns change. Compare El Niño conditions in the western and central tropical Pacific with the neutral conditions of Figure 1. During neutral conditions, surface air pressure in the central Pacific is higher (accompanied by fair weather) than to the west. During El Niño, the surface air pressure to the west is higher than in the central Pacific. This reversal in the atmospheric pressure pattern, called the *Southern Oscillation*, was first studied in an attempt to explain monsoon failure and drought in India.

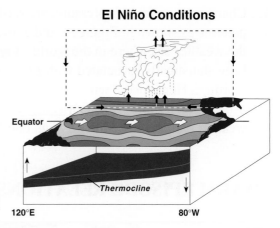

El Niño Conditions

Figure 2.
Diagram of El Niño Conditions in the tropical Pacific.

12. In response to changes in the air pressure pattern across the tropical Pacific, the trade winds weaken (and wind directions can reverse, especially in the western Pacific as shown by the bold dark arrows). No longer being pushed toward and piled up in the western Pacific, the warm surface water reverses flow direction. As shown by the surface currents arrows, the surface water during *El Niño* flows toward the east. As evident in the appropriate sea surface temperature shading, this causes SST in the eastern tropical Pacific to be **[(*higher*)(*lower*)]** than during neutral conditions.

13. In response to changes in surface currents, sea surface heights in the eastern tropical Pacific are higher than during neutral conditions. At the same time, the arrival of the warmer water in the east causes the surface warm-water layer to thicken. Evidence of this is the **[(*shallower*)(*deeper*)]** depth of the thermocline to the east compared with neutral conditions.

Tropical Pacific During *La Niña*

14. **Figure 3** shows atmospheric and oceanic conditions during *La Niña*. The tropical Pacific at times experiences trade winds stronger than neutral conditions with SST lower than usual in the eastern tropical Pacific and higher than usual in the western tropical Pacific. Because stronger trade winds produce stronger surface currents during *La Niña*, the warm water is pushed westward and colder water wells up to cause below-average sea-surface temperatures in the eastern tropical Pacific. It also follows that sea surface temperature in the western tropical Pacific must be **[(*above*)(*below*)]** its neutral condition average.

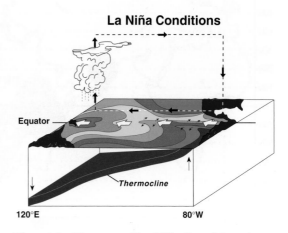

La Niña Conditions

Figure 3. Diagram of La Niña Conditions in the tropical Pacific.

15. Changes in surface air pressure, areas of large-scale convection, and upper air flow patterns as shown in Figures 2 and 3 alter the planetary wind circulation and affect the weather elsewhere in the world. **Figure 4** shows some weather patterns that have been statistically associated with *El Niño* conditions. This figure shows that during our Northern Hemisphere winter when *El Niño* is taking place, the southeastern states are usually **[(*drier and warmer*)(*wetter and cooler*)]** than normal. **Figure 5** shows some weather patterns linked to *La Niña* conditions.

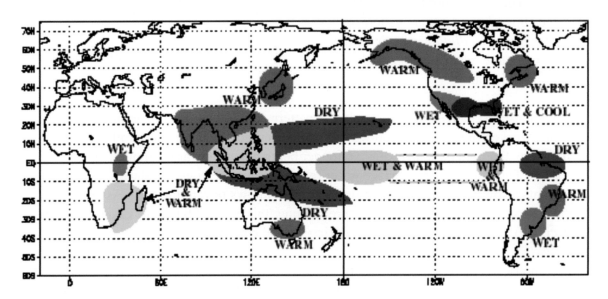

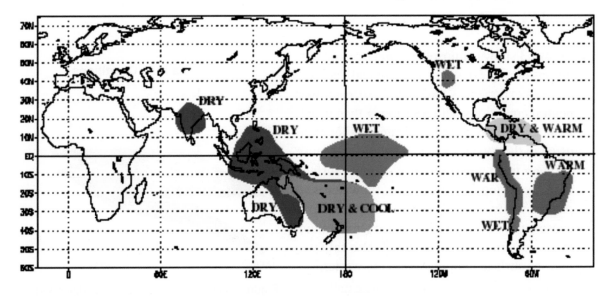

Figure 4. Weather patterns statistically associated with El Niño conditions.

COLD EPISODE RELATIONSHIPS DECEMBER - FEBRUARY

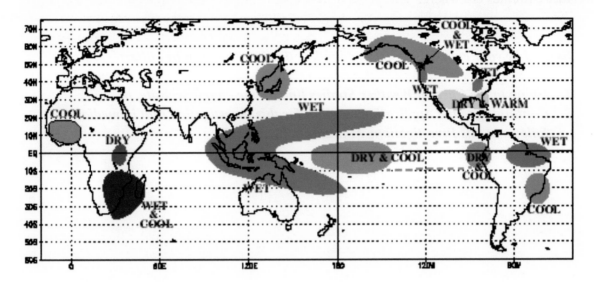

COLD EPISODE RELATIONSHIPS   JUNE - AUGUST

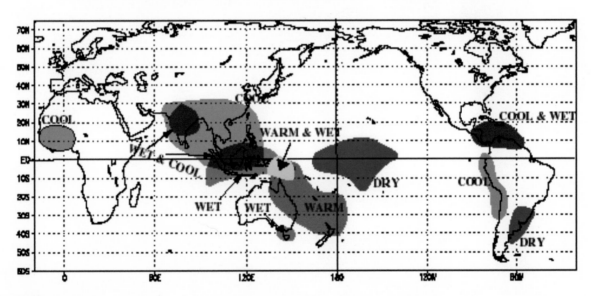

Figure 5. Weather patterns statistically associated with La Niña conditions.

The planetary-scale circulation of the atmosphere along the Intertropical Convergence Zone (ITCZ) includes the northeasterly trade winds of the Northern Hemisphere converging with the southeasterly trades of the Southern Hemisphere. But this generalized picture does not describe all the fluctuations of the dynamic Earth-atmosphere system. Changing temperatures in the upper layers of the Pacific Ocean and the overlying atmosphere along the equator lead to the Southern Oscillation and El Niño/La Niña episodes. In much of 1997 and early 1998, for example, the tropical Pacific Ocean was experiencing an unusually strong *El Niño*. The effects of these tropical ocean-atmosphere conditions extended well beyond the tropics and may well have set the stage for the extensive storminess along the West Coast, relatively warm and dry weather in the Southeast, the mild winter in the northern states, and weather extremes elsewhere.

As directed by your course instructor, complete this investigation by either:

1. *Going to the Current Weather Studies link on the course website, or*
2. *Continuing to the Applications section for this investigation that immediately follows in this Investigations Manual.*

Investigation 9B: **Applications**

¡EL NIÑO!

Following the intense El Niño episode of 1982-83 with its worldwide weather impacts, an instrumented array of buoys (Tropical Atmosphere Ocean (TAO) or TAO/TRITON array) was deployed across the tropical Pacific from ten degrees North latitude to ten degrees South latitude. **Figure 6** is a map showing the buoy locations. This array, along with satellite observations, has allowed realtime monitoring of tropical Pacific ocean and atmosphere conditions and provides input for models used to predict future episodes.

16. A representative reporting of TAO surface data is presented in **Figure 7**. The upper panel of Figure 7 depicts the five-day mean tropical Pacific sea surface temperatures (SST) and wind conditions ending on April 6, 2010. The SST are shaded with isotherms drawn at one-half degree intervals. Winds are shown by arrows. The shading and isotherms indicate that the warmest waters across the tropical Pacific were located from about **[(*160° E to 180°*)(*170° W to 150° W*)(*140° W to 120° W*)]** longitude. [Note, the Pacific east of 180° longitude has **W**(est) numbered longitudes while the Pacific west of 180° has **E**(ast) numbers.]

17. The winds across the tropical Pacific were generally from **[(*west to east*)(*east to west*)]**.

18. The lower panel of Figure 7 displays *Anomalies*, that is, departures from the long-term average. Positive temperature anomaly isotherms are solid lines, negative anomaly isotherms would be dashed lines. The interval between lines is one-half degree Celsius. A bold solid line denotes the 0-degree departure (*i.e.* average). The broad pattern of current SST anomalies shows **[(*overall negative*)(*overall positive*)(*negative in the west*)**

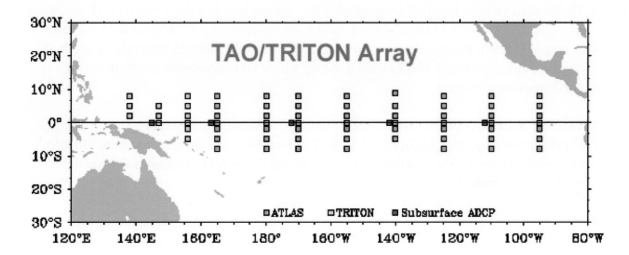

Figure 6.
The locations of TAO-Triton instrumented buoys in the tropical Pacific Ocean.
[*http://www.pmel.noaa.gov/tao/proj_over/map_array.html*]

***and positive in the east*)(*positive in the west and negative in the east*)]** values over the tropical Pacific region between 5 °N and 5 °S latitude.

19. The magnitude of the greatest positive anomalies is greater than **[(*0.5*)(*1.0*)(*1.5*)]** Celsius degrees.

20. The anomalous wind directions, i.e. departures from the average, are generally **[(*from the west*)(*from the east*)(*neither strongly west nor east*)]** over the tropical Pacific.

For contrast and comparison, we will look at TAO data acquired during recent significant El Niño and La Niña events. In much of 1997 and early 1998, the tropical Pacific Ocean experienced a strong El Niño. **Figure 8** is a depiction of the average ocean surface temperatures and atmospheric surface winds in the tropics over the month of November 1997 as measured by the TAO array, near the peak of the 1997-98 El Niño episode.

21. The top view (November 1997 Means) is the average sea surface temperatures and surface winds for the month of November 1997. The sea surface temperatures (SSTs) across the region ranged from about 26 °C as the "coolest" in the southeast corner to about 30 °C as the "warmest" just south of the Equator, west of center in the upper panel. These highest SSTs were located at about **[(*170° W*)(*120° W*)]** Longitude in the tropical Pacific.

22. The wind directions in the eastern Pacific were generally from the southeast. In the western Pacific, near the Equator (from about 140° E to 150° W), winds were generally light with some blowing from the west and some from the east. These observed winds and SSTs generally **[(*were*)(*were not*)]** consistent with the depiction of the Figure 2 schematic above for an El Niño. (For larger views of these schematics, see *http://www.pmel.noaa.gov/tao/elnino/nino_normal.html*.)

23. The bottom view of Figure 8 (November 1997 Anomalies) is a depiction of SST and wind *anomalies*, departures of the observed values shown in the top view from the long-term average. (Recall: Positive temperature anomalies are solid lines in intervals of one-half degree Celsius. A heavy line labeled **0** shows where no temperature anomaly exists, *i.e.* conditions are average.)

The SST anomalies in the eastern Pacific were positive, with the greatest values being slightly more than **[(*1.5*)(*4.5*)(*7.5*)]** C°. SST anomalies along the equator were virtually all positive or zero. The location and degree of the warm SST anomalies is what defines the El Niño situation.

24. Now examine **Figure 9**. These are the tropical Pacific SST and wind conditions for November 1998, one year later than Figure 8, showing that La Niña conditions had replaced El Niño. For November 1998, the sea-surface temperatures along the Equator in the eastern Pacific were near 22 °C, several degrees **[(*warmer*)(*cooler*)]** than those of the same area during the El Niño in November 1997. The winds across the entire Pacific area

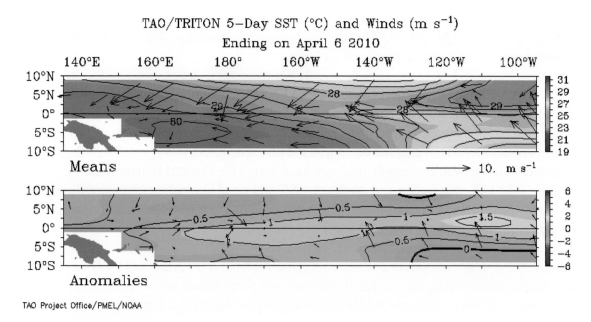

Figure 7. Five-day mean tropical Pacific sea surface temperatures and wind conditions ending on April 6, 2010.

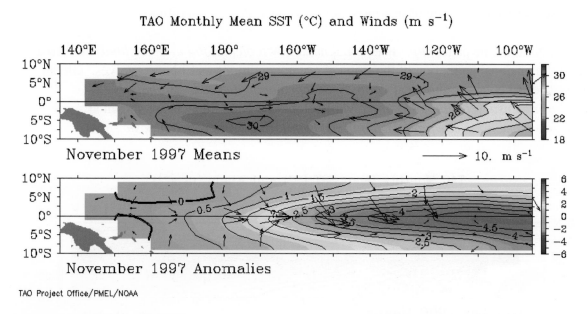

Figure 8. November 1997 oceanic means and anomaly conditions from TAO array.

of the depiction were now generally blowing from the east. The warmest waters were found in the extreme western Pacific.

25. These observed winds and SSTs in November 1998 generally [(*were*)(*were not*)] consistent with the depiction of those of this investigation's Figure 3 schematic for La Niña.

26. The lower panel of November 1998 *Anomalies* shows the Pacific SST anomalies

along the Equator being almost all negative, denoted by the dashed lines, with values dropping below [(**–2**)(**–3**)] C°. This relatively cool [compared to the Neutral ("Normal") Conditions] water is characteristic of La Niña.

27. The NOAA Climate Prediction Center (CPC) ENSO description of current Pacific conditions and their forecast can be found at *http://www.cpc.ncep.noaa.gov/products /analysis_monitoring/enso_advisory/ensodisc.html*. For several months prior to Figure 7, warm anomalies had covered the equatorial Pacific. The forecast suggested that those warm conditions would strengthen somewhat through the fall and remain over the winter. The CPC discussion and the warm SST and wind anomalies (compare Figure 7 to Figure 8) across the central tropical Pacific are all consistent with the existence of [(***a La Niña***)(***neutral conditions***)(***an El Niño***)].

For additional displays of current Pacific information related to El Niño and La Niña conditions, including SSTs, anomalies, depth cross-sections, winds and some animations, go to: *http://www.cpc.ncep.noaa.gov/products/precip/CWlink/MJO/enso.shtml*.

Suggestions for further activities: You might investigate the El Niño/La Niña websites given above to determine the instrumentation used to obtain these *in situ* oceanic buoy measurements. Also, the sites display the Southern Oscillation Index (SOI). You can explore the discovery and meaning of this indicator of tropical Pacific conditions.

The El Niño theme page, *http://www.pmel.noaa.gov/tao/elnino/nino-home.html*, links to a three-dimensional animation of the tropical ocean conditions as El Niño evolves. Global impacts of El Niño are shown at: *http://www.cpc.noaa.gov/products/analysis_ monitoring/ensocycle/elninosfc.shtml*, while global La Niña impacts are shown at: *http://www.cpc.noaa.gov/products/analysis_monitoring/ensocycle/laninasfc.shtml*.

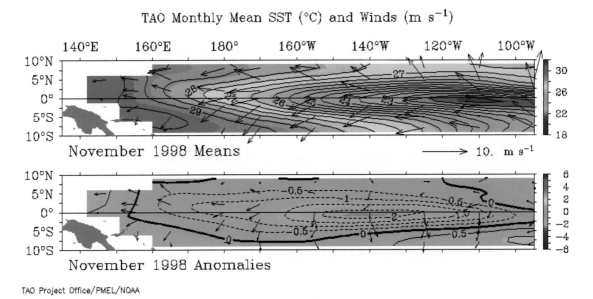

Figure 9. November 1998 oceanic means and anomaly conditions from TAO array.

Investigation 10A:

THE EXTRATROPICAL CYCLONE

Objectives:

Extratropical cyclones are storm systems of the middle and high latitudes characterized by low-pressure centers, fronts, and variable weather conditions. The counterclockwise and inward circulation that characterizes surface winds in Northern Hemisphere cyclones brings contrasting air masses together to form fronts. Along those fronts clouds and precipitation may develop. As a cyclone travels along its track, the system typically progresses through a life cycle. Localities that come under the influence of a cyclone often experience sequential changes in weather. Understanding the types of weather associated with a typical extratropical cyclone aids in weather forecasting.

After completing this investigation, you should be able to:

- Describe the pattern of surface winds and weather in a model extratropical cyclone.
- Specify the type of weather associated with fronts that rotate about an extratropical cyclone's low-pressure center.
- Compare and contrast the weather associated with cold fronts and warm fronts.

Introduction:

Figure 1 is a schematic of a Northern Hemisphere extratropical cyclone reaching maturity. North is to the top and East is to the right. Shown are the positions of the cold and warm fronts along with isobars drawn at intervals of 4 mb. **Label the fronts with the appropriate symbols for a warm front and cold front. Draw arrows to show wind directions to the northeast, southeast, southwest, and northwest of the cyclone center.**

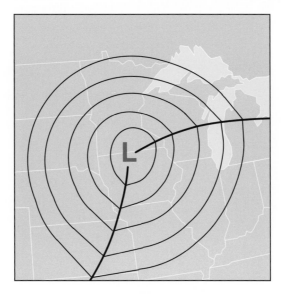

Figure 1. Schematic diagram of a Northern Hemisphere extratropical cyclone.

Refer to Figure 1 and prior course information when responding to the following:

1. Looking down on a Northern Hemisphere extratropical cyclone, surface winds blow [(*__clockwise and outward__*)(*__counterclockwise and inward__*)] about the center.

2. The specific track of an extratropical cyclone's low-pressure center across Earth's surface is largely determined by large-scale horizontal winds blowing [(*__near Earth's surface__*)(*__in the middle and upper troposphere__*)].

3. As a cyclone moves forward, the system typically progresses through a life cycle. As a cyclone develops, the central pressure of the system [(*__falls__*)(*__rises__*)] and surface winds strengthen. At maturity, clouds cover a broad area about the low center and associated precipitation is widespread.

4. In the extratropical cyclone's warm sector (the area between the warm front and the advancing cold front), surface winds are likely to be producing [(*__warm__*)(*__cold__*)] air advection.

5. To the north and west of the center of an extratropical cyclone, surface winds are likely to be producing [(*__warm__*)(*__cold__*)] air advection.

6. Dewpoints are likely to be relatively high to the [(*__southeast__*)(*__northwest__*)] of the cyclone's center.

7. As the cyclone progresses across Earth's surface, the cold and warm fronts rotate about the center of low pressure. The motion of the storm system has similarities to that of a flying Frisbee®, that is, a Frisbee spins as it sails through the air (simultaneously exhibiting rotational and translational motions). Typically, the cold front rotates about the center of the low faster than the warm front. Hence, eventually the cold front catches up with and merges with the warm front forming an occluded front. At this stage in the life cycle of an extratropical cyclone (known as *occlusion*), the storm often begins to weaken as the central air pressure begins to [(*__rise__*)(*__fall__*)].

8. As a cyclone progresses through its life cycle, its cold front rotates toward its warm front. The extent of the warm sector occupied by relatively warm and humid air at the surface [(*__shrinks__*)(*__increases__*)].

9. With the passage of a warm front, the air temperature usually rises and the dewpoint usually [(*__falls__*)(*__rises__*)].

10. With the passage of a cold front, the air temperature usually falls and the dewpoint usually [(*__falls__*)(*__rises__*)].

11. A shift in wind direction usually accompanies the passage of a front. With passage of the cold front, surface winds shift direction from the south to the [(*__southeast or east__*)(*__west or northwest__*)].

12. With passage of the warm front, surface winds shift direction from the east to the [(*southeast or south*)(*west or northwest*)].

13. Ahead of a surface warm front, warm and humid air rides up and over a wedge of cooler air (a process known as *overrunning*.) As the ascending warm air expands and cools, its relative humidity [(*increases*)(*decreases*)], and clouds typically form.

14. Most cloudiness associated with a warm front develops over a broad area, often hundreds of kilometers wide, [(*ahead of*)(*behind*)] the front. From these clouds, light to moderate precipitation may fall for 12 to 24 hours or longer.

15. As a cold air mass advances and a warm air mass retreats, the colder, denser air forces the warmer, lighter air to ascend either along or just ahead of the cold front. Uplift of warm air triggers cloud development and perhaps showery precipitation. In some instances, uplift is so vigorous that thunderstorms develop. Typically, the band of clouds and precipitation associated with a cold front is [(*narrower*)(*wider*)] than that associated with a warm front.

16. **Figure 2** is a visible satellite image for 1815Z 23 APR 2010. The frontal positions at that time have been added to the satellite view to characterize the structure of a mature extratropical cyclone evidenced on a surface map for the same time. The storm system had experienced an occlusion.

    Around the main low-pressure system centered in eastern Colorado, the visible image shows a broad white, comma shaped swirl of clouds. Consistent with the hand-twist model of a Low, an animation would show this swirl to be rotating [(*clockwise*)(*counterclockwise*)]. It is the circulation and rising motions of low-pressure weather systems that leads to the "comma" shape of cloudiness frequently seen in satellite images.

17. A broad, bright, "bumpy" white patch of clouds over Louisiana, Arkansas and Mississippi is the result of extensive precipitation from thunderstorms at image time. These thunderstorms are broadly spread ahead of the wave cyclone's [(*warm*)(*cold*)] front.

While extratropical cyclones typically exhibit similar general characteristics, they do not all look alike or go through exactly the same life cycles. The "open wave" stage of the extratropical cyclone seen in Figure 2 occurred some time prior to this view. The map one and one-half days previously showed a developing low-pressure center in Utah with short warm and cold fronts that began the development of the cyclonic circulation.

As directed by your course instructor, complete this investigation by either:

1. *Going to the Current Weather Studies link on the course website, or*
2. *Continuing to the Applications section for this investigation that immediately follows in this Investigations Manual.*

Visible Image   1815Z   23 APR 2010

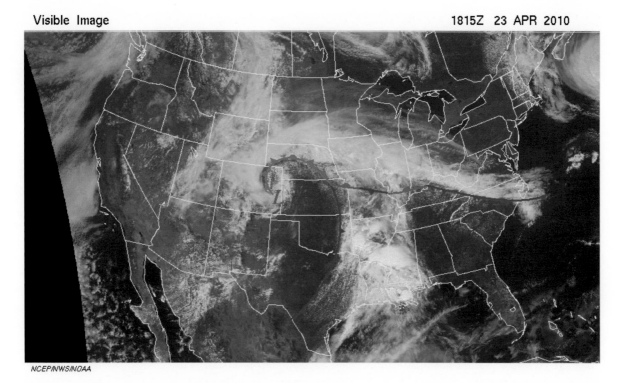

NCEP/NWS/NOAA

Figure 2. Visible satellite image for 1815Z 23 APR 2010.

Investigation 10A: Applications

THE EXTRATROPICAL CYCLONE

In early April 2010, a strong low-pressure system had developed to the lee (east side) of the Rockies along a nearly stationary cold front. Subsequently, it moved northeastward across the central U.S. This investigation looks at the structure of that well-developed midlatitude extratropical cyclone.

18. **Figure 3** is the surface weather map for 12Z 08 APR 2010. At map time, the center of the low-pressure system was shown by an *L* within the closed 996-mb isobar, located over eastern Michigan. From that L, a [(*cold*)(*warm*)(*stationary*)] front extended to the southeast into western New York State. The front continued generally eastward into the Atlantic as a stationary front.

19. From the L, a [(*cold*)(*warm*)(*stationary*)] front stretched southward and then southwest-ward into the Gulf and finally northern Mexico.

20. The wind pattern in the region impacted by the low-pressure area centered on the *L* exhibited a generally [(*clockwise and outward*)(*counterclockwise and inward*)] flow as expected. This is also consistent with the directional symbols of the advancing fronts.

21. The winds from Wisconsin to eastern Texas were generally from the [(*southwest*)(*northwest*)(*northeast*)] as would be expected behind (following) the cold front.

22. The winds from Florida to Pennsylvania in the wave cyclone's warm sector (between the warm and cold fronts) were generally from the [(*south*)(*southwest*)(*northwest*)].

23. Compare the air temperature and dewpoint (32 °F and 30 °F, respectively) at Green Bay, WI behind the cold front and Low center with those at Albany, NY in the cool air north of the frontal boundary marking the leading edge of the warm sector. The air was warmer and more humid [(*east*)(*west*)] of the Low center.

24. Compare the air temperature and dewpoint at Atlanta, GA in the warm sector with those at Little Rock, AR behind the cold front. The air was warmer and more humid [(*in the warm sector between the fronts*)(*behind the cold front*)].

25. The air was generally least humid [(*north of the warm/stationary front*)(*in the warm sector between the fronts*)(*north and west of the cold front*)].

26. The area of more-intense precipitation associated with this midlatitude cyclone, as represented by the radar echo shadings on the map, was located primarily [(*to the east of the Low center*)(*in a broad band along the cold front*)]. The area of precipitation northwest of the Low center was falling as snow, denoted (faintly) by three stars for the present weather condition at Green Bay.

27. **Figure 4** is the 500-mb upper-air map for 12Z 08 APR 2010, the same time as the surface map. Place a bold "**L**" on the Figure 4, 500-mb map where the surface Low center is shown. The center of the storm system at the surface, as marked by the L from the Figure 3 map, is located to the east of the axis of a weak upper air [(*ridge*)(*trough*)] shown on the 500-mb map over the north-central U.S.

28. Based on the wind directions at 500-mb over the area of the surface storm center, the system would be expected to move generally toward the [(*southeast*)(*northeast*)] over the next day or so with its trailing front moving generally eastward.

29. As a result, persons in states along the East Coast could expect generally [(*cooler*) (*warmer*)] temperatures as the frontal system advances eastward over the next day or so.

30. **Figure 5** is the surface weather map for 12Z 09 APR 2010, twenty-four hours after the Figure 3 map. At this time the central pressure of the Low was 1000 mb, higher than the approximately 992 mb at 12Z on the 8th. The cyclone was therefore weakening, exhibiting less strong horizontal pressure gradients, and undergoing the process of becoming a mature midlatitude cyclonic system. This was most evident with the presence of a(n) [(*cold*)(*warm*)(*occluded*)] front that had formed extending from the Low center in Quebec Province of Canada to the triple point where the three fronts met at the *L* shown in Connecticut. This front is shown in purple with alternating semicircles and triangles on the same side of the front.

31. The cold front in Figure 5 is shown located just offshore from Connecticut to where it crossed the central Florida peninsula and then extended into the eastern Gulf of Mexico. Note the temperatures and winds at Greensboro, NC. Over the twenty-four hour period between maps, the wind flow at Greensboro went from southerly to northwesterly and the temperatures at Greensboro had become [(*cooler*)(*warmer*)], as expected.

These maps also show a midlatitude cyclonic system along the West Coast, to the east of another upper air trough, which traveled into the mountains and became occluded. The NWS National Centers for Environmental Prediction (NCEP) also creates surface maps that can be chosen for North America, the continental U.S., Alaska, Hawaii or sections of the country, in either black and white or color, with or without fronts, and/or satellite or radar. They are found from the "NWS Surface Analyses" link on the website under **Surface** products. Also, one- and seven-day loops of frontal positions can be created showing the progression of surface weather features. The National Weather Service Unified Surface Analysis will even show (and animate) wave cyclones and Highs across most of the Northern Hemisphere.

Suggestions for further activities: NOAA weather radio, "State Surface Data - Text" on the course website, or local instrument readings and media reports are ways you can keep track of changes in hourly weather conditions that accompany cyclones as these systems and their fronts approach and cross your location, particularly when severe weather threatens. You can record these hourly weather conditions on a blank meteogram, available from the course website, under **Extras**, "Blank Metgram". Then you can compare these changing

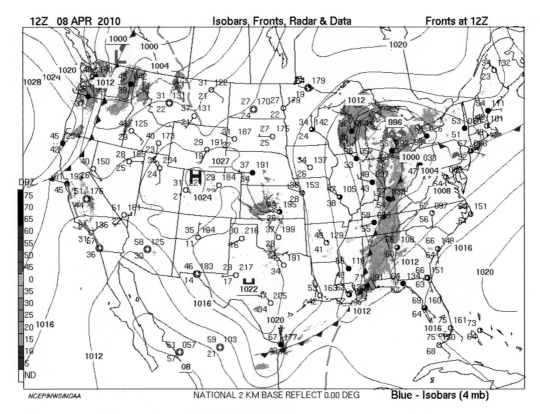

Figure 3. Surface weather map for 12Z 08 APR 2010.

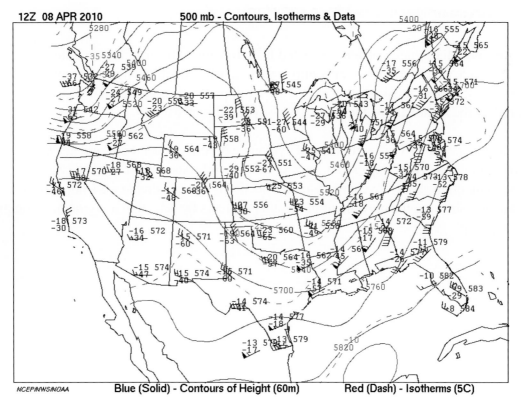

Figure 4. 500-mb map for 12Z 08 APR 2010.

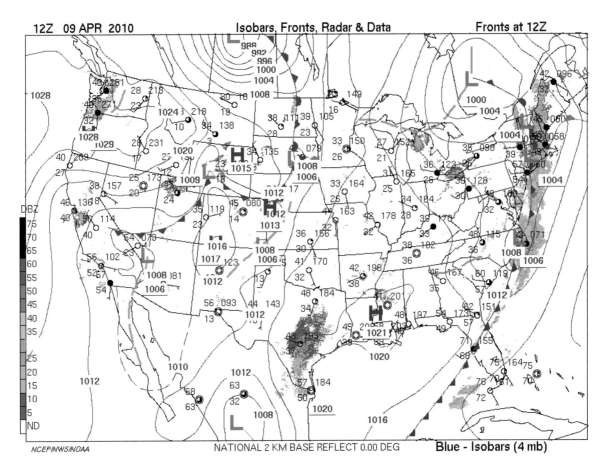

Figure 5. Surface weather map for 12Z 09 APR 2010.

conditions with the general pattern expected from passing extra-tropical cyclones (recall Investigation 5A). NOAA Weather Radio is particularly valuable when severe weather threatens your area as most weather radios are equipped with warning alarms that can be triggered by the National Weather Service when conditions warrant.

Additional discussion of the extratropical cyclone with images can be found at: *http:// ww2010.atmos.uiuc.edu/(Gh)/guides/mtr/cyc/home.rxml*. Also, from the *NWS Surface Analyses* link on the course website (**Surface** section), you can create a 24-hour surface analysis loop showing the movement of weather systems by selecting a region and clicking "Display Loop".

EXTRATROPICAL CYCLONE TRACK WEATHER

Objectives:

An understanding of the typical weather patterns associated with a typical extratropical cyclone enables us to forecast probable changes in weather as the cyclone approaches and then moves away from a particular location, including where we live. As a general rule, the weather on one side of a storm track differs from the weather on the other side. The storm track is the path the low-pressure center takes as the system progresses through its life cycle. Facing in the direction towards which a Northern Hemisphere cyclone is moving, the weather is usually colder on the left side of the storm track and warmer on the right side of the storm track. Typically, only localities on the right side of a storm track experience the passage of fronts. During the colder times of the year, the storm track can separate areas of snow from areas of rain.

After completing this investigation, you should be able to:

- Describe the sequence of changes in weather that typically takes place on the right (warm) side of a cyclone track.
- Describe the sequence of changes in weather that usually takes place on the left (cold) side of a cyclone track.

Introduction:

1. **Figure 1** shows a winter low-pressure system that at map time is intensifying over eastern Colorado. Over the subsequent two days, the storm system is forecast to track towards the Great Lakes region. Track A would have the cyclone center passing by Detroit to the southeast of the city and Track B would have the cyclone center passing by to the northwest of Detroit. In either case, the system would pass close enough to Detroit to have major consequences for the city's weather. The potential positions of the cyclone center along each of the two tracks (Track A and Track B) at 12-hour intervals are indicated by heavy dots. As the cyclone tracks toward the Great Lakes, the system progresses through its life cycle. The cold and warm fronts gradually rotate (counterclockwise as viewed from above) about the cyclone center with the faster cold front closing in on the slower warm front.

   Using Track A frontal positions shown in Texas as the starting point, pencil in the cyclone's estimated cold and warm frontal positions at 12-hour intervals along its predicted path. Draw the fronts from the low-pressure center at each location. For simplicity, draw the warm front in essentially the same position relative to the low pressure center and gradually rotate the cold front until it is oriented roughly north/ south 24 hours after its Texas position. With Track A, residents of Detroit [(**_do_**)(**_do not_**)] experience the passage of fronts.

2. Apply the *hand-twist model* of low pressure systems to the cyclone's position at 12-hour intervals along Track A. Assume that before the storm's arrival the wind at Detroit is blowing from the east. As the cyclone approaches the wind shifts from the east to the [(**_southeast_**)(**_northeast_**)].

3. Considering the wind shifts and frontal positions at Detroit as the cyclone passes through the region along Track A, the city is on the relatively [(**_warm_**)(**_cold_**)] side of the system.

4. **Using the cyclone's Track B frontal positions in Oklahoma as a guide, pencil in the associated cold front and warm front at 12-hour intervals.** Follow the same guidelines you employed in drawing Track A frontal positions. With Track B, residents of Detroit [(**_do_**)(**_do not_**)] experience the passage of fronts.

5. **Apply the *hand-twist model* of low pressure systems to each of the storm's 12-hour positions along Track B.** Assume that initially the wind at Detroit is blowing from the east. As the storm approaches the wind shifts from the east to the [(**_southeast_**)(**_northeast_**)].

6. Considering the wind shifts at Detroit as the cyclone's center passes through Michigan along Track B, Detroit is positioned to experience weather related to the relatively [(**_warm_**)(**_cold_**)] side of the system.

7. One extended period of substantial snowfall at Detroit is more likely if the cyclone takes Track [(**_A_**)(**_B_**)].

8. Two periods of precipitation, more likely to be rain, separated by a short period of relatively warm fair weather at Detroit might occur if the cyclone takes Track [(**_A_**)(**_B_**)].

9. As the cyclone center approaches Detroit on either storm track, the air pressure at the city [(**_falls_**)(**_rises_**)].

10. As the cyclone center moves away from Detroit, the air pressure at the city [(**_falls_**)(**_rises_**)].

11. The next weather system to affect Detroit is likely a cold [(**_cyclone_**)(**_anticyclone_**)] approaching from central Canada.

As directed by your course instructor, complete this investigation by either:

1. *Going to the Current Weather Studies link on the course website, or*
2. *Continuing to the Applications section for this investigation that immediately follows in this Investigations Manual.*

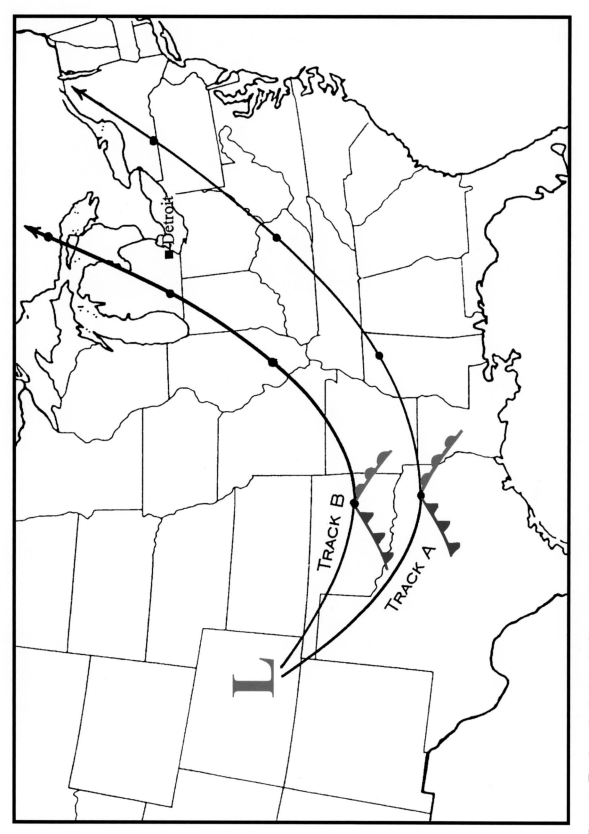

Figure 1. Two hypothetical tracks for an extratropical cyclone.

Investigation 10B: Applications

EXTRATROPICAL CYCLONE TRACK WEATHER

In this portion of the Investigation we consider further details of the path of the storm system we examined in Investigation 10A that crossed the central U.S. As the storm system traveled northeastward, cold air from Canada was drawn southward behind the low-pressure center. The clash of warm, humid air streams from the south mixing with cold, drier air streams from the north set the stage for the variety of precipitation types observed with this storm's track.

12. **Figure 2** is the surface weather map for 12Z 07 APR 2010. This was the display of weather conditions twenty-four hours prior to the Figure 3 map of Investigation 10A. At 12Z 07 APR 2010, the primary low-pressure center was situated along the frontal system in northeast Missouri with a pressure of 999 mb (label appears north of the partially obscured L). The temperatures at stations from the Great Lakes to western Kansas where radar echo shadings indicated precipitation was occurring suggested that the precipitation was likely in the form of **[(*rain*)(*snow*)]**. Further west, the station model for Denver in north-central Colorado shows a temperature of 31 °F and two stars for the present weather condition signifying that snow was occurring at map time. (The *User's Guide* linked from the webpage displays common weather symbols.)

13. Review the Figure 3 and Figure 5 surface weather maps from Investigation 10A. On that Investigation 10A Figure 5 map of 12Z 09 APR, plot a bold *L* at the positions of the 12Z 07 APR Low center in northeast Missouri from this Figure 2 map and the 12Z 08 APR Low center in eastern Michigan (Inv. 10A Figure 3) on the final 12Z 09 APR map (Inv. 10A Fig. 5). Connect the three *L*s with a bold curved line to denote the direction of travel of the storm's low-pressure center, the "storm track". The storm center tracked generally toward the **[(*northeast*)(*northwest*)(*southeast*)(*southwest*)]**.

14. This direction **[(*was*)(*was not*)]** consistent with the direction expected from the 500-mb flow displayed on the Investigation 10A Figure 4 map (Inv. 10A question 28).

Figure 3 is a composite view from the NWS National Operational Hydrologic Remote Sensing Center of precipitation for the twenty-four hour period ending at 06Z on 9 April 2010 in the area near the western Great Lakes. This is substantially the period between the last two surface maps of the storm track series. The left display is of snowfall in inches. The display on the right is non-snow precipitation (essentially rain) also in inches. The respective precipitation are below each view. There is a shading scale for topographic elevation as well. Place a bold *L* in eastern Michigan on both views to denote the low-pressure center's position at 12Z on 8 April.

15. The extensive area of precipitation from the mid-Mississippi River Valley to the Great Lakes generally **[(*does*)(*does not*)]** coincide with the area crossed by the track of the storm center and the advancing fronts.

16. The greatest precipitation amounts were shaded purple to red where values are about one inch liquid-equivalent amounts. These greater precipitation amounts were in the non-snow view and were located generally from east of Lake Michigan to Ohio. These greater rates likely [(*__would__*)(*__would not__*)] have accompanied thunderstorm and rain activity.

17. This non-snow pattern seemed to "wrap around" the L marked in eastern Michigan. This circulation pattern would have brought warmer air from the south into the storm's low-pressure circulation and promoted precipitation that was likely [(*__snow__*)(*__rain__*)].

18. In the left, snow view, the snow area was located well to the [(*__northwest__*)(*__southeast__*)] of the marked Low center. It is seen that the greatest snow amounts were located to the lee (downwind) sides of eastern Lake Superior and along Green Bay of Lake Michigan where the winds may have produced lake-effect snows or mixtures of snow and rain.

19. During the storm's passage, the "cold" side would be to the [(*__left__*)(*__right__*)] looking along the directional arrows you placed on the Investigation 10A Figure 5 map (storm track).

20. During the storm's passage, the "warm" side would be to the [(*__left__*)(*__right__*)] looking along the directional arrows you placed on the Investigation 10A Figure 5 map (storm track).

21. These relative positions of the respective precipitation types in this actual storm [(*__are__*) (*__are not__*)] consistent with the model presented in the introductory portion of Investigation 10B.

Many surface map products can be found from the course website's **Surface** section, "NWS Surface Analyses" link. Links near the top of NCEP's North American Surface Analysis page provide another archive of surface weather maps for the past two years. Links further down the page provide surface maps for regions in a variety of formats as well as animations for the past day at three-hourly intervals. There is also a link at the bottom of the **NWS Surface Analyses** page to the official how-to book, *NWS Unified Surface Analysis Manual*.

Figure 3 is from *http://www.nohrsc.noaa.gov/nsa/* where snow depths and other snow parameters for national and regional areas of various periods can be found. They also animate images so an entire season of snowfall can be relived! Another asset is the NWS Advanced Hydrologic Prediction Service (*http://water.weather.gov/*) where national, regional and state precipitation maps for periods of the previous one day, one or several weeks or season as well as archived times can be found.

An additional valuable Internet weather resource is NOAA's Daily Weather Map series. This website allows one to find historical maps of detailed surface and 500-mb conditions at 7 AM EST (12Z) on any date from 1 September 2002 onward. Also shown are maps of daily maximum and minimum temperatures and total precipitation for that day. Go to the Daily Weather Map series website at *http://www.hpc.ncep.noaa.gov/dailywxmap/*. The map displayed is the most recent posted, typically a day or two in the past. The menu in the frame

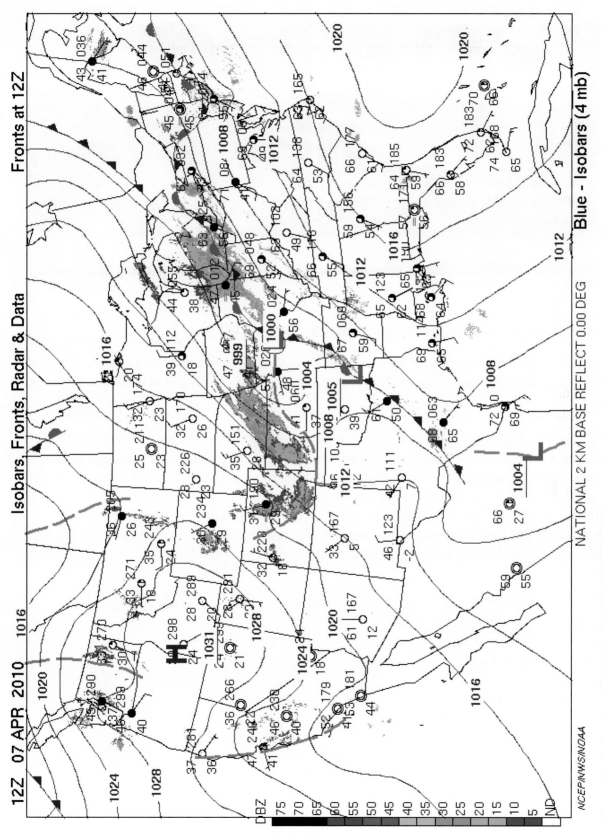

Figure 2. Surface weather map for 12Z 07 APR 2010.

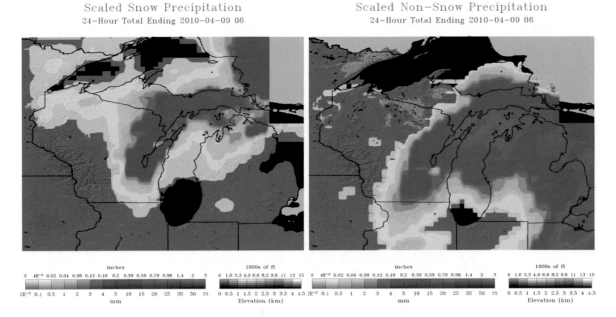

Figure 3.
Accumulated 24-hour snowfall (left) and rainfall (right) ending at 06Z 09 APR 2010.

to the left allows one to choose the date for prior maps. The top panel lets you choose any of the last 16 days posted as a quick pick. The second panel provides a map date selection back to 2002.

Suggestions for further activities: Actual weather forecasts issued to the public by NOAA's National Weather Service are produced by meteorologists beginning with computer guidance information. You might use the website forecast maps to compare the NWS forecast weather conditions for your location with the conditions that actually occur at the forecast "Valid Time." Forecasts issued by local radio and television stations or newspapers are usually based on NWS forecasts. (Generally, weather forecasts appearing in newspapers are the least accurate due to the long lead time needed to meet publication deadlines.)

Objectives:

A thunderstorm is one of nature's most awesome spectacles; and it is a major mechanism whereby heat energy is transported from Earth's surface into the atmosphere. Thunderstorms are also responsible for dangerous lightning, intense rainfall that can lead to flash flooding, damaging surface winds, and tornadoes. Thunderstorms are the consequence of convection currents that surge to great altitudes within the troposphere (and sometimes into the lower stratosphere). At middle latitudes, surface heating by the Sun can set the stage for thunderstorm development. However, convection forced by converging surface winds or uplift along a frontal surface or a mountain slope produces most midlatitude thunderstorms.

After completing this investigation, you should be able to:

- Describe the appearance of thunderstorms on visible satellite imagery.
- Identify probable locations of thunderstorms on infrared satellite imagery.
- List some of the modes of occurrence of thunderstorms.

Introduction:

1. As a general rule of thumb, the greater the altitude of the top of a thunderstorm cloud (cumulonimbus), the more intense the thunderstorm cell. A relatively high thunderstorm top implies vigorous convection and a relatively **[(_weak_)(_strong_)]** updraft.

2. Within a thunderstorm cell, the temperature **[(_falls_)(_rises_)]** with increasing altitude primarily because of the expansion of rising air within the cloud.

3. An intense thunderstorm thus has a relatively **[(_cold_)(_warm_)]** cloud top.

4. On a visible satellite image, a large thunderstorm can appear as a bright white blotch, or cluster. The brightness of the blotch indicates that the cloud top has a relatively **[(_high_) (_low_)]** albedo for visible solar radiation.

As directed by your course instructor, complete this investigation by either:

1. ***Going to the Current Weather Studies link on the course website, or***
2. ***Continuing to the Applications section for this investigation that immediately follows in this Investigations Manual.***

Investigation 11A: Applications

THUNDERSTORMS

In Investigations 10A and 10B we examined the case of a strong cyclonic midlatitude storm system. Investigation 10B, Applications section, showed the types of precipitation involved with the cold and warm sides of that storm track. We continue here with that system looking at the precipitation derived from thunderstorms that occurred along the warm side of the track.

Figure 1 is the surface weather map for 00Z 09 APR 2010 (Thursday evening, April 8th). [This map showed weather conditions 12 hours prior to Figure 5 of Investigation 10A.] In Figure 1, the storm system's low-pressure center was located over Canada's southeastern Ontario Province with an occluded front to New York State where it met the warm and cold fronts. The cold front extended southwestward to the Florida panhandle and Gulf of Mexico.

5. Note the temperature and dewpoint at Jacksonville, in northeastern Florida located in the warm sector of the storm system and at Mobile, Alabama behind the cold front. Jacksonville's temperature and dewpoint were 77 °F and 59 °F, respectively while Mobile's were 68 °F and 37 °F. The air in the warm sector of the system was generally [(*cool and dry*)(*warm and humid*)] compared to that over the Southeast following the front.

6. The cold front was moving generally toward the [(*northeast*)(*southeast*)(*southwest*)].

7. The radar echoes denoting precipitation were generally located [(*well behind*)(*just ahead of and along*)] the cold front.

8. Bright red specks within the radar shading areas, particularly in a long, narrow band from northern Florida to central New York, denote very heavy rainfall intensities. This [(*is*)(*is not*)] consistent with the existence of thunderstorms.

9. The dewpoints across the Florida and Carolinas area, ahead of the cold front, [(*do*)(*do not*)] indicate that there was abundant near-surface moisture for precipitation formation.

10. The direction of movement of the cold front toward the East Coast area [(*does*)(*does not*)] indicate that there was a lifting mechanism for that humid air. At two places ahead of the front, the notation "OUTFLOW BOUNDARY" has been added. This indicates that strong gusty winds which often accompany the strong downdrafts of thunderstorms were likely occurring and local weather radars were displaying patterns signifying possibly damaging winds.

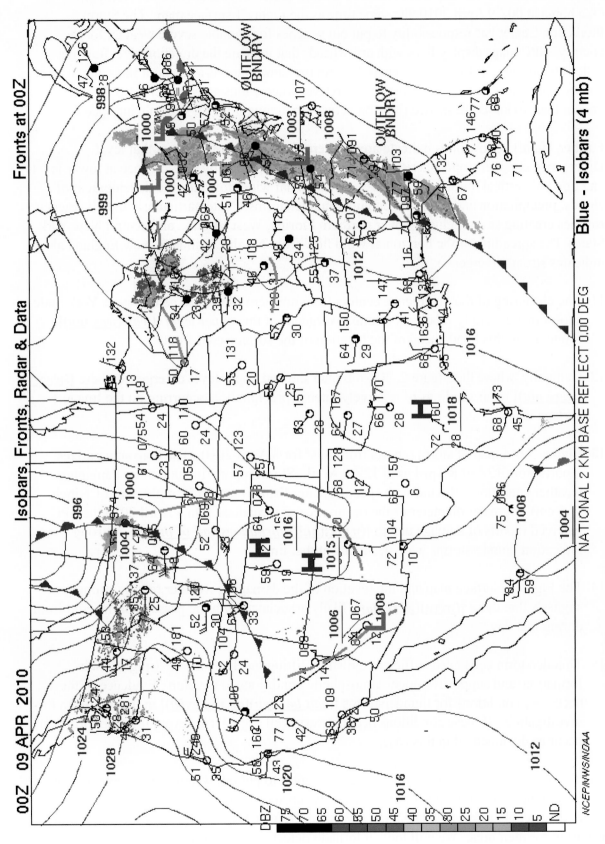

Figure 1. Surface weather map for 00Z 09 APR 2010.

Figure 2 is the 300-mb map from NOAA's Storm Prediction Center (SPC) in Norman, Oklahoma at 00Z 9 April 2010, the same time as the Figure 1 surface map. The Storm Prediction Center has responsibility to put out watches for possible severe weather for the country. SPC maps display lines with arrowheads that indicate the direction air is flowing, called streamlines. (Streamlines generally form the same patterns as contour lines drawn on a constant pressure-chart for the same pressure level when winds are strong.) Also the isotachs greater than 70 knots are shaded (blue on-screen) to highlight the jet stream wind speeds.

Further, (faint) yellow lines indicate areas where *divergence* is present at that level. Divergence at upper tropospheric levels must be accompanied by convergence at lower levels and <u>upward</u> vertical motions. This three-dimensional wind flow aids the development of clouds, precipitation and, if strong enough, thunderstorms. Note particularly the divergence isolines creating several clusters over northern Florida, West Virginia and central New York State. The spreading of the horizontal wind flow from North Carolina to New England also indicates strong divergence.

11. The clustering of the divergence centers from northern Florida to central New York State, shows that generally relatively strong divergence at this upper level [(***was***)(***was not***)] found along this band ahead of the location of the surface cold front.

12. Generally, where the Figure 2, 300-mb map showed upper level divergence, there [(***were***) (***were not***)] radar shadings of relatively intense precipitation occurring on the Image 1 surface map.

13. **Figure 3** is a map of storm reports from SPC for 8 April 2010. The map summarizes reports from 12Z of 8 April until 12Z 9 April. Storm reports include those damaging weather conditions of wind with speeds greater than 50 knots (58 mph), hail of 1-inch (2.5 cm) or greater diameter or the occurrence of tornadoes. On 8 April 2010 the SPC received reports of damage due to [(***wind***)(***hail***)(***tornadoes***)(***all of these***)], a strong indication thunderstorms were occurring at that time.

14. The Figure 1 surface map and the location of the damaging weather reports strongly suggests that these [(***were***)(***were not***)] likely associated with the passage of the cold front during that day.

15. Thunderstorm cells need sufficient moisture, lifting mechanisms to "trigger" their formation, and supportive upper atmospheric conditions to develop. Evidence of these necessary conditions for thunderstorms [(***can be seen***)(***do not exist***)] in the figures in this investigation. If these conditions become very well organized, severe thunderstorms can occur as documented in this case.

It is the responsibility of the Storm Prediction Center to put out weather **watches** for such severe weather conditions for the entire U.S. When such conditions are actually occurring or are detected by radar, a local National Weather Service Office puts out a specific weather **warning** for their area. We will look at tornado detection and warning in Investigation 11B.

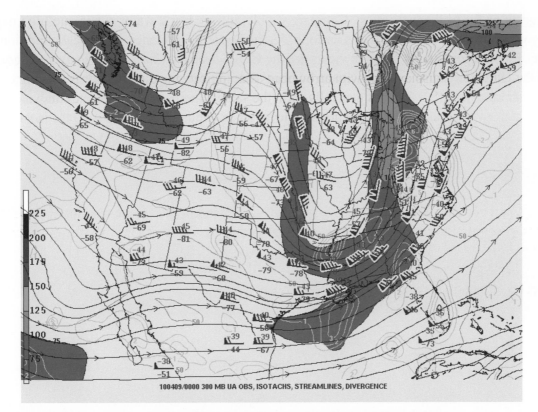

Figure 2. 300 mb constant pressure map for 00Z 09 APR 2010. [NOAA SPC]

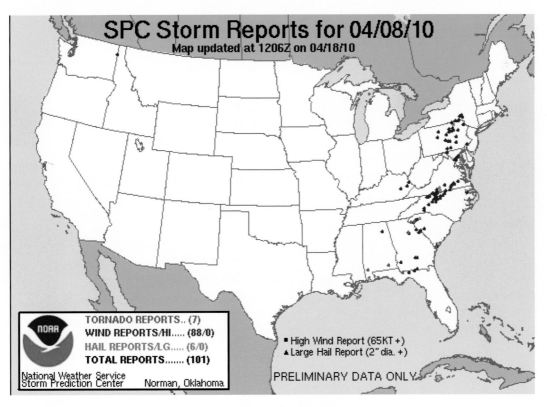

Figure 3. Map of storm reports for 8 APR 2010. [NOAA SPC]

Suggestions for further activities: Compare visible, infrared, and water vapor satellite images for your region with radar views as various weather systems influence your local conditions. Other Internet sources for satellite views are the GOES website (*http://www.goes.noaa.gov/*) and the University of Wisconsin, Space Science and Engineering Center (*http://www.ssec.wisc.edu/data/*).

Investigation 11B:

TORNADOES

Objectives:

A tornado can be the most intensely violent weather system on Earth. Its extreme winds can take lives and cause considerable property damage. Wind speed in a tornado, estimated from the extent of property damage, is the basis for rating these systems from 0 (weakest) to 5 (strongest) on the new Enhanced Fujita scale (EF-scale). Most tornadoes are spawned by and travel with severe thunderstorms and generally exhibit a counterclockwise rotation as seen from above. The special weather pattern required for tornadic thunderstorms to develop is most common in spring and summer in the central United States.

After completing this investigation, you should be able to:

- List some of the characteristics of the path of an intense tornado.
- Describe the general weather conditions favorable for formation of tornadic thunderstorms.
- Explain why winds on one side of a tornado may be stronger than winds on the other side.

Introduction:

In the early evening of 3 May 1999, an outbreak of severe weather struck the southern Great Plains. Many tornadoes touched down in central Oklahoma, but the most devastating twister tracked through the suburbs and portions of downtown Oklahoma City.

A portion of the National Weather Service Public Information statement on this tornado read as follows:

"JUST SOUTH AND EAST OF AMBER... THE TORNADO QUICKLY GREW TO CLOSE TO THREE-QUARTERS OF A MILE WIDE. ASPHALT PAVEMENT... ABOUT ONE INCH THICK... WAS PEELED FROM A SECTION OF RURAL ROAD /EW125 RD/ ABOUT FIVE MILES EAST OF STATE ROAD 92. THE FIRST DAMAGE RATED AT F4 WAS DISCOVERED ABOUT FOUR MILES EAST-NORTHEAST OF AMBER. F4 DAMAGE WAS OBSERVED CONTINUOUSLY FOR SIX AND ONE-HALF MILES... WITH ANOTHER AREA OF F4 DAMAGE ABOUT 2 MILES NORTHWEST OF NEWCASTLE. TWO AREAS OF F5 DAMAGE WERE OBSERVED. THE FIRST WAS IN THE WILLOW CREEK ESTATES... A RURAL SUBDIVISION OF MOBILE HOMES AND SOME CONCRETE SLAB HOMES IN BRIDGE CREEK. TWO HOMES WERE FOUND COMPLETELY SWEPT FROM THEIR SLABS... AND ABOUT ONE DOZEN AUTOMOBILES WERE CARRIED ABOUT ONE-QUARTER MILE. GRASS VEGETATION IN THIS AREA WAS COMPLETELY SCOURED TO MUD... AND SMALL CEDAR TREES WERE LEFT DE-BARKED AND DEVOID OF GREENERY. THE RIDGECREST BAPTIST CHURCH WAS DESTROYED NORTHEAST OF THE FIRST F5 DAMAGE AREA.

THE SECOND F5 DAMAGE WAS ONE MILE WEST OF THE COUNTY LINE IN BRIDGE CREEK...
AND CONSISTED OF A CLEANLY SWEPT SLAB HOME WITH FOUNDATION ANCHOR BOLTS AND
ANOTHER VEHICLE LOFTED ONE-QUARTER MILE. THE MAXIMUM WIDTH OF THE TORNADO
IN BRIDGE CREEK WAS ABOUT ONE-MILE WIDE. THE TORNADO MAINTAINED A NEARLY
STRAIGHT PATH TO THE NORTHWEST OF THE TURNPIKE EXCEPT WHEN IT MADE A SLIGHT
JOG TO THE RIGHT AND MOVED DIRECTLY OVER THE 16TH STREET TURNPIKE OVERPASS
BEFORE RESUMING ITS ORIGINAL COURSE. THE TORNADO CONTINUED INTO THE NORTHERN
SECTIONS OF RURAL NEWCASTLE... AND CROSSED THE TURNPIKE AGAIN JUST NORTH OF THE
U.S. 62 NEWCASTLE INTERCHANGE. AT THIS LOCATION THE TORNADO NARROWED TO ABOUT
ONE-QUARTER MILE WIDE AND THE DAMAGE INTENSITY DROPPED TO F2 BEFORE IT CROSSED
THE SOUTH CANADIAN RIVER INTO NORTHERN CLEVELAND COUNTY."

Preliminary reports indicated 44 fatalities and over 700 injuries with this tornado.
Approximately 11,000 homes were destroyed and total insured damage was about $750
million. The destruction throughout central Oklahoma from this outbreak was estimated to
total about $1.2 billion.

1. As shown on the damage paths map in **Figure 1** (north is at the top of the map), the
   tornadoes in this outbreak generally traveled **[(*toward the northeast*)(*toward the
   southeast*)]**. This is the general direction of movement of the majority of tornadoes that
   occur in the United States. [Map courtesy of NOAA's NWS, Norman, OK.]

2. The most devastating tornado in this outbreak was the one that tracked through Oklahoma
   City (center of map, continuous path beginning south of Amber, designated F5). The first
   reports of tornado damage along the track that led into Oklahoma City were at 7:17 PM
   CDT when multiple injuries occurred. Damage near the end of the path was reported at
   8:10 PM CDT. Hence, this tornado was on the ground for at least **[(*17*)(*43*)(*53*)]** minutes.

3. Based on the Figure 1 map scale, the length of this tornado path on the ground was about
   [(*20*)(*40*)(*80*)] miles.

4. Therefore, this tornado's forward speed averaged about **[(*25*)(*45*)(*85*)]** miles per hour.

5. Numbers associated with the tornado paths on the map are F-scale values. The "**F**"
   designations for tornadoes refer to levels of destruction based on the Fujita Tornado
   Intensity Scale. The Fujita scale and associated wind speeds are shown in the upper
   left corner of the map (These are original Fujita-Scale values that were in effect prior
   to 1 February 2007). The highest F-scale shown on the map with the Oklahoma City
   tornadoes was **[(*F-2*)(*F-3*)(*F-4*)(*F-5*)]**.

6. Tornadoes are classified by the highest reported F-scale rating as assessed by related
   structural damage. The "5" value was shown in purple numbers at **[(*1*)(*3*)(*8*)]** point(s)
   along the damage path of the tornado we have been examining.

7. According to the Fujita scale presented in Figure 1, a tornado with this rating is
   accompanied by maximum wind speeds estimated to be within the range of **[(*158-206*)
   (*207-260*)(*261-318*)(*over 318*)]** mi. per hour.

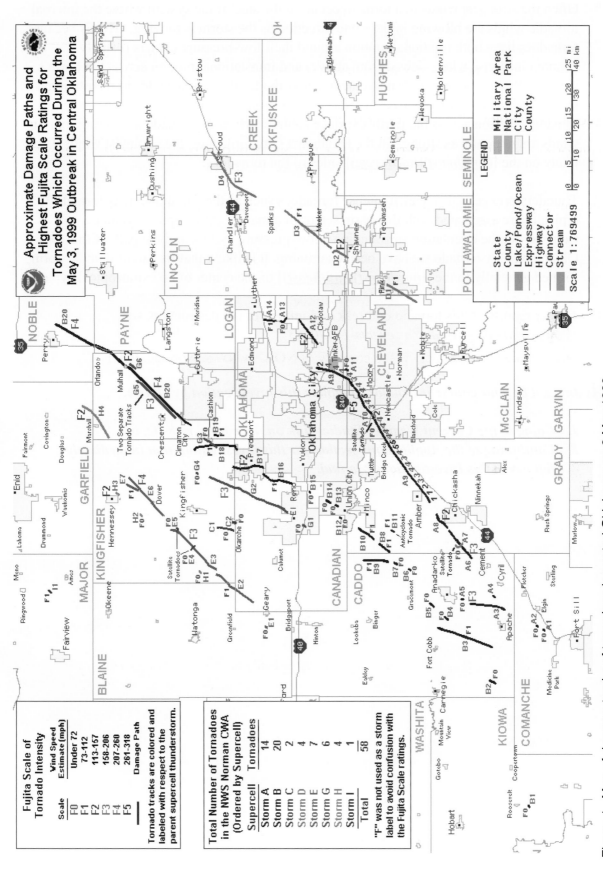

Figure 1. Map of damage tracks of tornadoes in central Oklahoma on 3 May 1999.

8. Often the strongest winds in a tornado occur on the side of the system where the internal tornadic winds are blowing in the same direction as the storm is moving. This is because tornadoes have both rotational motion around their low-pressure centers (usually counterclockwise when viewed from above) and translational motion across the Earth's surface. Where the tornado's movement across the Earth's surface adds to its rotational motion, winds relative to the Earth's surface are strongest. On the other side where the tornado's motion across the Earth's surface subtracts from the tornado's spinning motion, winds are not quite as strong. In the Oklahoma City tornadoes, the strongest winds were likely on the [(*southeast*)(*northwest*)] sides of the tornadoes.

9. Thunderstorm cells that spawn intense tornadoes such as the Oklahoma City tornado typically are associated with the warm sector of a mature extratropical cyclone. Considering the circulation surrounding a low-pressure system and the path of the Oklahoma City tornado we have been examining, it is likely that its parent thunderstorm formed generally to the [(*northwest*)(*southeast*)] of the center of an extratropical cyclone.

For further reports of this tornado outbreak and images of the destruction, see:
http://www.srh.noaa.gov/oun/?n=events-19990503

As directed by your course instructor, complete this investigation by either:

1. *Going to the Current Weather Studies link on the course website, or*
2. *Continuing to the Applications section for this investigation that immediately follows in this Investigations Manual.*

Investigation 11B: Applications

TORNADOES

The 2010 severe weather season in terms of tornadoes had been relatively quiet until the latter half of April. Spring is typically the most active tornado season, with April to June having the greatest numbers and severity of tornado occurrence. Until the last third of April, only 15 tornadoes had been reported. January, February and March tallied 28, 1 and 36, respectively, versus the latest three-year averages for those months of 37, 78 and 138. In fact, the single tornado in February tied a record for the fewest ever reported in that month. Also, there had been only one fatality, significantly under the most recent 3-year average of 14 during the January to April period. Among recent years, 2008 had 1691 tornadoes with 549 occurring during the first four months, causing 70 deaths.

Suddenly, a single storm system that formed on April 22nd and moved slowly across the central U.S. turned violent and spawned incredible amounts of severe activity. From 22 to 24 April, the thunderstorm activity with this system formed 136 tornadoes that resulted in ten deaths. One spectacularly deadly and devastating tornado occurred on 24 April and traversed the Southeast from Louisiana to Mississippi for two hours wreaking destruction reminiscent of the infamous Tri-State tornado of 1925. We will examine the deadly tornado's impact in western Mississippi on 24 April.

The NOAA Storm Prediction Center (SPC) is one of NOAA's National Centers for Environmental Prediction. Their website is at *http://www.spc.noaa.gov/*. The SPC provides severe thunderstorm and tornado *watches* for the coterminous U.S. (The *warning* of an actual impending or occurring event is issued by the local National Weather Service office.)

10. **Figure 2** is the SPC map of severe weather reports: tornadoes, damaging winds and hail for the twenty-four hour period 12Z (7 am CDT) 24 April to 12Z 25 April 2010. The number of tornado reports on this single day was **[(*many fewer than*)(*about the same as*) (*many greater than*)]** the total number reported for the first 3.5 months of 2010.

11. A long streak of red triangle tornado reports extends from northeast Louisiana to southwestern North Carolina. Earlier damage reports were toward the southwestern end of the streak while later reports occurred toward the northeastern end. This trend also was seen in other severe weather reports that day. Such general southwest to northeast movement of intense thunderstorm cells **[(*would*)(*would not*)]** be expected to be consistent with the general flow of middle tropospheric winds.

12. **Figure 3** is the surface weather map for Saturday morning, 12Z 24 APR 2010. This map shows the weather conditions across the U.S. several hours prior to the formation of the deadly tornado that later struck Yazoo City, Mississippi. At map time Jackson, MS, located to the southeast of Yazoo City had a temperature of 74 °F and a dewpoint of 72 °F. The wind at Jackson was shown as being from the **[(*north*)(*east*)(*south*)(*west*)]** at 10 knots.

13. This flow of humid air from the Gulf [(*would*)(*would not*)] have provided the inflow of moisture, one necessary ingredient for thunderstorm development.

14. Along the Louisiana-Texas border to the west was a complex low-pressure center from which trailed a [(*warm*)(*cold*)(*occluded*)] front.

15. The frontal system and the low-pressure system advancing eastward [(*would*)(*would not*)] have provided a lifting mechanism, another necessary ingredient for thunderstorm development.

Figure 4 is the 300-mb map from NOAA's Storm Prediction Center (SPC) at 12Z 24 April 2010, the same time as the Figure 3 surface map. SPC 300-mb maps display lines with arrowheads that indicate the direction air is flowing, called streamlines and isotachs with areas of winds greater than 70 knots shaded in blue to highlight jet stream wind speeds. Yellow lines indicate areas where *divergence* is present at that level. Divergence at upper tropospheric levels provides the uplift to support rapidly rising motions essential to severe thunderstorms.

16. Note particularly the divergence isolines creating several clusters from southern Missouri to the Gulf, particularly over northeast Louisiana and northern Mississippi. The spreading of the wind flow across this region also indicates strong divergence. These indicators of upper-level instability [(*were*)(*were not*)] the third necessary ingredient for thunderstorm development. The intensity of this upper-level support was instrumental in SPC issuing a tornado watch #91 for the northeast LA and northern MS region from 6:00 am until 1:00 pm CDT.

Figure 5 is the map of tornado tracks determined from surveys of the damage paths of tornadoes in the northeast LA and northern MS area on 23 and 24 April developed by the NWS Forecast Office in Jackson, MS. While there are seven tracks shown, we want to concentrate on the main track shown in red across the center of the map with the large information box highlighting that track. This Yazoo City tornado was determined to have initially set down 5 miles west of Tallulah in Madison Parish, LA at 11:06 am CDT and dissipated 5.5 miles north of Sturgis in Oktibbeha County (east of Choctaw), MS at 1:52 pm.

The "vital statistics" associated with the Yazoo City tornado were:

Total time:	1 hr 46 min.
EF scale rating:	4
Path length:	149.25 mi.
Path width:	1.75 mi.
Est. ground speed:	84 mph

17. On the Figure 5 track map, north is at the top of the map. The Yazoo City tornado moved in the direction toward the [(*southeast*)(*southwest*)(*northeast*)(*northwest*)]. This direction, accompanying thunderstorms embedded in mid-tropospheric winds, is typical of most tornadoes in the U.S.

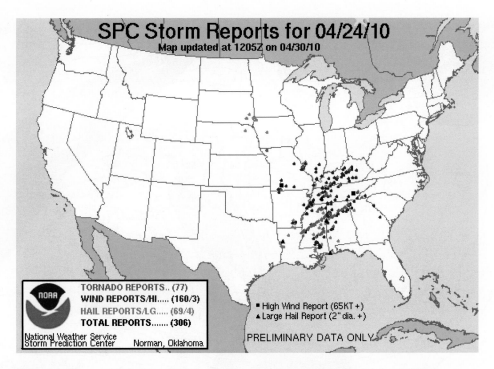

Figure 2. NOAA Storm Prediction Center severe weather reports for 24 April 2010.

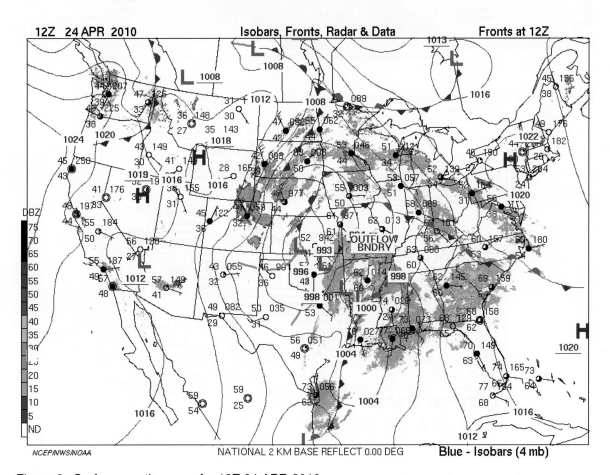

Figure 3. Surface weather map for 12Z 24 APR 2010.

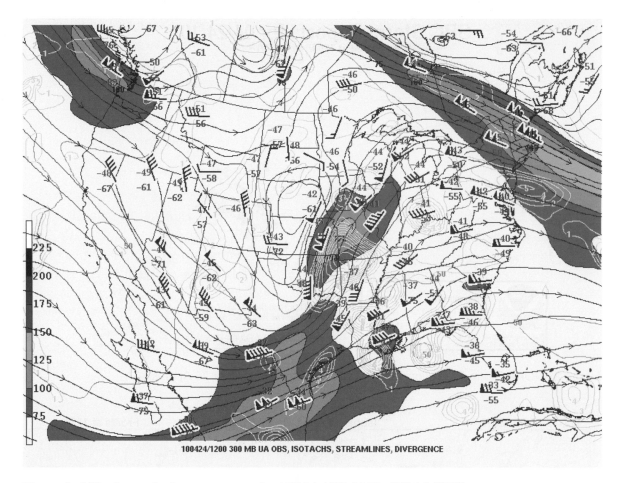

Figure 4. 300 mb constant pressure map for 12Z 24 APR 2010. [NOAA SPC]

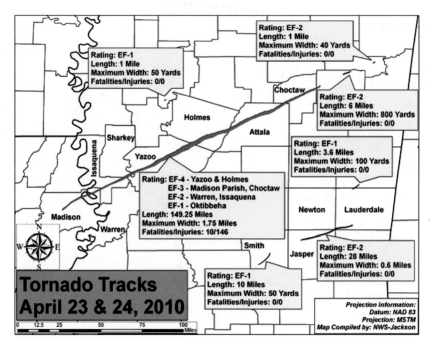

Figure 5. Tornado damage path tracks across northwestern Mississippi on 23-24 April 2010. [NWS Jackson]

The *Enhanced Fujita (EF) Scale* (*http://www.depts.ttu.edu/weweb/EFScale.pdf*) with
equivalent winds speeds of 3-second gusts is given in **Table 1**. [Note that this EF Scale
is different from the F Scale in effect when the 3 May 1999 tornado outbreak in central
Oklahoma (Figure 1) occurred. When examining tornado damage reports, determine which
scale is employed.]

Table 1. Enhanced Fujita (EF) Scale Wind Speed Ranges

EF Scale	3-Second Gust Speed (mph)
EF 0	65 - 85
EF 1	86 - 109
EF 2	110 - 137
EF 3	138 - 167
EF 4	168 - 199
EF 5	> 200

18. Based on the EF category determination for the Yazoo City tornado, wind speeds could
    have ranged from [(**_65-85_**)(**_110-137_**)(**_168-199_**)] mph. An EF-4 category tornado would
    produce devastating damage to structures and create deadly missiles of detached objects.

19. Based on the speed of travel of the tornado, it [(**_would_**)(**_would not_**)] be wise to try to
    outrun such a threat in a vehicle. With such power and widespread destruction path, only
    the safety afforded by a small, reinforced, ground (or lower) level, window-less room is
    likely to be sufficient.

20. **Figure 6** is the composite view of the radar reflectivity on the left and the radial velocity
    on the right from the Jackson NWS office at 1714Z (12:14 pm CDT) on 24 April. Yazoo
    city is located at the images' center. The shadings of the left reflectivity view show the
    thunderstorm cell over Yazoo City. The large curl of continuous bright red shading from
    the image top to Yazoo City is the "hook echo" often seen in tornadic thunderstorms.
    Also, the dark red oval, intense reflectivities, just southeast of Yazoo City was likely the
    result of debris in the air from ruined structures. These shadings [(**_would_**)(**_would not_**)]
    suggest tornadic activity to be a possibility near Yazoo City.

21. The right storm relative radial velocity view in Figure 6 displays the *tornadic vortex
    signature (TVS)* of red and blue colors adjacent to each other southeast of Yazoo City.
    The radar site is located to the southeast of Yazoo City, beyond the lower right corner of
    the view. Recall from Investigation 7B that red denotes Doppler velocities <u>away</u> from
    the radar site and blue or green is <u>toward</u>. Draw a short arrow <u>away</u> from the radar site
    across the brightest red section of the pixels. Also draw a short arrow <u>toward</u> the radar
    site across the bright blue portion. These represent the radial velocities away and toward
    the radar's location. Your pattern of arrows suggests a circulation that is [(**_clockwise_**)
    (**_counterclockwise_**)].

The full report of the Yazoo City tornado is at *http://www.srh.noaa.gov/jan/?n=2010_04_23_24_tor_outbreak_front_page*. The webpages provide descriptions of the weather situation, display several radar reflectivity and relative velocity images and several photos of the damage.

More information on the Enhanced Fujita Scale is available at: *http://www.spc.noaa.gov/efscale/*. The first occurrence of an EF-5 tornado was in Greenberg, KS in May 2007 (*http://www.crh.noaa.gov/news/display_cmsstory.php?wfo=ddc&storyid=9475&source=2* and *http://www.crh.noaa.gov/images/ddc/News/Greensburg/Greensburg_1year_later.pdf*).

Additional reports of other tornado episodes containing NWS radar imagery can be accessed at the following links:
- An April 2008 episode of several deadly tornadoes and other wind damage in Mississippi: *http://www.srh.noaa.gov/jan/?n=2008_04_04_svr*.
- Western Tennessee tornadoes on 2 April 2006 including many photos of property damage can be found at: *http://www.srh.noaa.gov/meg/?n=apr22006tornadooutbreak*.
- The Evansville tornado of November 2005 can be found at: *http://www.crh.noaa.gov/pah/?n=evansvilletornado-nov.6,2005*. Included are the tornado track, radar images and many damage pictures.
- The central Iowa tornadoes of November 12, 2005: *http://www.crh.noaa.gov/dmx/?n=stratford-woodward-11122005*.
- LaPlata, MD, Apr. 28, 2003 unusually devastating East Coast tornado: *http://www.erh.noaa.gov/er/lwx/Historic_Events/apr28-2002/laplata.htm*
- The Xenia, OH, September, 2000 tornado that almost repeated 1974, *http://www.erh.noaa.gov/er/iln/92000.htm*.

Your local NOAA/NWS office websites may have links to notable severe weather episodes in your area. Finally, for an account of the historic Super Tornado Outbreak of 1974, see: *http://www.publicaffairs.noaa.gov/storms/*. For more on the infamous Tri-State Tornado of 1925, see: *http://www.crh.noaa.gov/pah/1925/*.

Last, but not least, a site to answer (almost?) all of your tornado questions: *http://www.spc.noaa.gov/faq/tornado/*.

Suggestions for further activities: Investigate the Internet tornado pages given in this investigation. Examine the types of radar imagery available to forecasters to use in issuing severe weather and tornado warnings. Current radar imagery, including single station views (NEXRAD), is available from the "NWS Radar Page" link on the course website. Selecting a location on the interactive map will allow one to see regional views and select individual station reports. Views of reflectivity in lowest level scan ("base") and greatest intensity of any level ("composite") are available along with base and storm relative velocities and 1-hour and storm total precipitation amounts. These can also be animated.

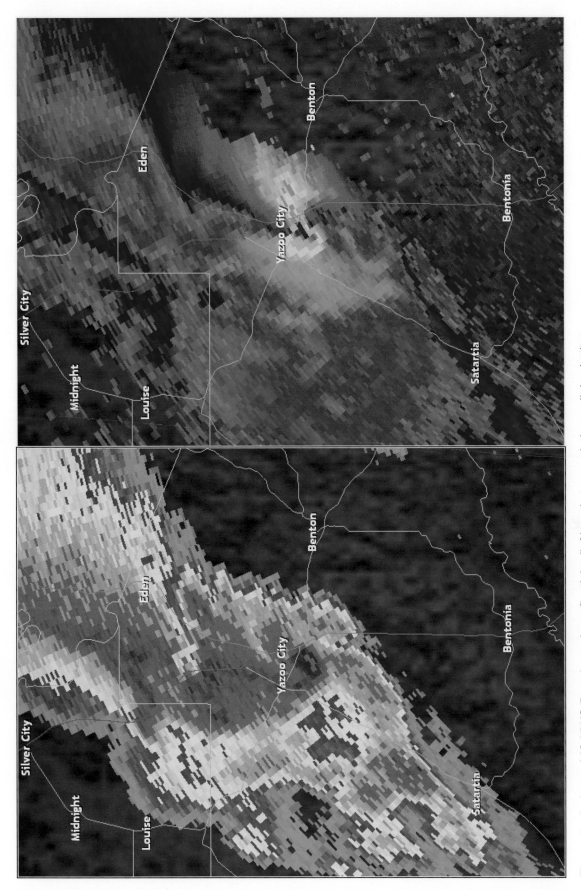

Figure 6. Jackson, MS NWS Doppler radar reflectivity (left) and storm relative radial velocity

HURRICANES

Objectives:

A hurricane is a tropical cyclone that has maximum sustained surface wind speeds of 119 km per hour (74 mph, 64 kt) or higher. [A knot (kt) is one nautical mile per hour.] A hurricane forms over the warm tropical ocean and derives its energy from latent heat released when water evaporated from the sea surface condenses in the storm system. A typical hurricane is about one-third the size of a typical extratropical cyclone of the middle latitudes, forms in a uniform mass of warm and humid air, and has no fronts or frontal weather. When a hurricane strikes the coast, property damage is caused by a surge of ocean water above flood stage, strong winds, heavy rainfall, and sometimes tornadoes.

Hurricanes that threaten the East and Gulf Coasts of North America usually originate over the tropical North Atlantic off the West African coast, the Caribbean Sea, or the Gulf of Mexico. Most hurricanes initially are steered slowly westward by the trade winds, but eventually curve northwestward, then northward, and finally northeastward around the Bermuda-Azores semi-permanent subtropical High. Precisely where the curvature takes place determines whether the hurricane strikes the Gulf Coast, the East Coast, or turns out to sea. However, a hurricane may depart significantly from this "average" track. In some cases, a hurricane meanders about, even moving in circles or figure-eights. Such behavior greatly complicates the task of hurricane forecasting.

After completing this investigation, you should be able to:

- Describe the track taken by a hurricane that occurred in the western North Atlantic Ocean.
- Indicate the probable position of highest storm surge when a hurricane makes landfall.

Introduction:

The 2005 Hurricane Season in the Atlantic Basin (including the Caribbean and Gulf of Mexico) was a record-breaker. In terms of the number of tropical cyclones, there were 27 named storms (tropical storms or hurricanes) compared to the long-term average of about 10 and the previous record of 21 in 1933. Fifteen of the tropical cyclones reached hurricane intensity, which is nine above average. And seven of the hurricanes were major systems (Saffir-Simpson category 3 or higher) versus an average of three. The major story, of course, was that four of those hurricanes reached the maximum intensity of category 5 and three made landfall along the U.S. Gulf Coast (*Katrina*, *Rita*, and *Wilma*).

Responsibility for the forecasting and warning of tropical weather systems in the Atlantic and eastern portion of the Pacific Ocean basins resides with the National Hurricane Center (NHC) in Miami, FL. The NHC's website: *http://www.nhc.noaa.gov/*, contains the latest information on tropical weather systems as well as a wealth of historical and other information regarding hurricanes.

This investigation involves evaluating the forecast track of an intensifying tropical cyclone that began as a tropical depression and quickly reached tropical storm strength. It formed over the southern Bahamas at the beginning of the observation period. This storm was Katrina.

"Katrina was an extraordinarily powerful and deadly hurricane that carved a wide swath of catastrophic damage and inflicted large loss of life. It was the costliest and one of the five deadliest hurricanes to ever strike the United States. . . . Considering the scope of its impacts, Katrina was one of the most devastating natural disasters in United States history." (*Tropical Cyclone Report, Hurricane Katrina, NHC, http://www.nhc.noaa.gov/2005atlan.shtml*)

The following **Figures 1**, **2**, and **3** are selected from the NHC Katrina Graphic Archives: *http://www.nhc.noaa.gov/archive/2005/KATRINA_graphics.shtml*. Each figure's inset legend shows information including the stage of development, date and time along with the sequential number of the advisory issued. Color coding shows coastal and land areas where watches or warnings had been issued, as described in **Table 1**:

Table 1. Tropical Storm and Hurricane Watches and Warnings

Color	Tropical weather statement	Expected wind speeds
yellow	tropical storm watch	34 - 63 kts (39-73 mph) within 36 hours
blue	tropical storm warning	34 - 63 kts (39-73 mph) within 24 hours
pink	hurricane watch	64 kts or greater (>74 mph) within 36 hours
red	hurricane warning	64 kts or greater (>74 mph) within 24 hours

The center of the circulation at that time was plotted in each figure with a dot in an orange circle. Additionally, large black dots display the forecast position centers with times showing expected tropical storm (**S**) or hurricane (**H**) strength and the white cone of forecast uncertainty in future positions.

We will examine three advisories issued by NHC as it tracked the storm that became Hurricane Katrina and impacted southern Florida and then went on to devastate so much of the Gulf Coast.

Figure 1 is the first advisory issued by the NHC when this tropical system reached enough strength and potential to merit attention. The advisory was communicated to emergency managers and the public at 5 PM Eastern Daylight Time on 23 August 2005. As described in the inset legend, the center of the system's circulation was located over the southern Bahamas at that time. The maximum sustained wind speed was determined to be 35 mph and the system was moving toward the northwest at 8 mph.

1. At 5 PM EDT on Tuesday, 23 August, the status of this tropical weather system, as described in the figure legend, was a [(***tropical depression***)(***tropical storm***)(***hurricane***)].

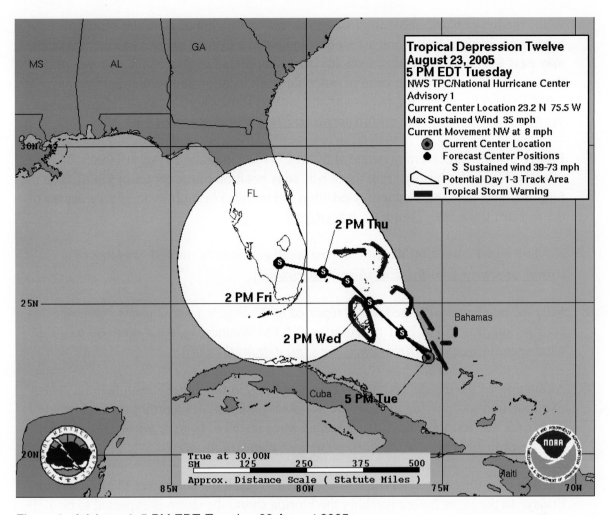

Figure 1. Advisory 1; 5 PM EDT, Tuesday, 23 August 2005.

2. The forecast path showed that the system was expected to travel generally northwestward through the Bahamas and make landfall on the Florida coast, at the northern edge of the Miami metropolitan area, at about midnight on Thursday. At the time of landfall in Florida, the system was expected to be a [(*tropical depression*)(*tropical storm*)(*hurricane*)].

3. Note that there are no colored shadings indicating watches or warnings shown along the Florida coastline. This time of predicted landfall on the Florida coast was about 55 hours beyond the advisory time. The definitions of watches and warnings given in Table 1 indicate that there [(*could*)(*could not*)] have been a watch or warning posted for this area at 5 PM on Tuesday.

4. If you lived in the threatened region of south Florida and knew of the counterclockwise circulation of the advancing storm as seen from above, you should expect the highest storm surge to occur to the [(*south*)(*north*)] of the point of the center's landfall (intersection of the heavy track line with the coast).

5. While the heavy black line is the forecasters' most probable track of the center of the storm's circulation, NHC forecast models indicated a distinct probability that the center may pass within the cone displayed, in decreasing probability to either side away from the center track. This white cone of potential track location shows that in three days (at 2 PM Friday) the center might possibly be located somewhere between [(*Georgia and the Mexican Yucatan Peninsula*)(*northern Florida and northern Cuba*)].

6. **Figure 2** is the forecast map issued at 5 PM EDT Wednesday, 24 August 2005, 24 hours after the Figure 1 advisory. Plot the Wednesday position of the center in Figure 2 on the Figure 1 map. Did the Wednesday position fall within a latitude or longitude degree of the forecast track from Tuesday? [(*Yes*)(*No*)].

7. As shown by Figure 2 on Wednesday, the weather system's strength was that of a [(*tropical depression*)(*tropical storm*)(*hurricane*)].

8. Note the maximum sustained winds reported in the Figure 1 and 2 insets. Over the 24-hour period, from 5 PM EDT on Tuesday to 5 PM Wednesday, the system (in terms of wind speed) [(*weakened*)(*remained the same*)(*strengthened*)].

9. At the time of the Figure 2 advisory, it was reported in the figure's inset that the center of circulation was moving towards the northwest at 9 mph. According to the forecast map, over the next three days, the system was expected to [(*curve toward the north*) (*continue straight toward the northwest*)(*curve toward the west*)].

10. The Figure 2 forecast track projects the system to reach the coast near Miami early on Friday. In the hours before reaching the Florida shore, the storm is expected to be a [(*tropical depression*)(*tropical storm*)(*hurricane*)].

11. Therefore, in its passage from the Bahamas to Florida, the system is expected to [(*weaken*)(*remain the same*)(*strengthen*)].

12. Following the system's landfall and its passage over the Florida peninsula, the strength of the system is projected to be that of a tropical storm (S). This is less than the strength at landfall. Weakening of the system's circulation is likely due to [(*loss of the latent heat source over water*)(*increased friction over land*)(*both of these factors*)].

13. Jumping forward two days to 5 PM EDT on Friday, 26 August 2005, **Figure 3** shows the storm positioned in the Gulf of Mexico. Plot the Friday position of the circulation center in Figure 3 on the Figure 2 map. Did the Friday position fall within a latitude or longitude degree of the forecast track from Wednesday? [(*Yes*)(*No*)].

14. At 5 PM EDT on Friday, the system was a [(*tropical depression*)(*tropical storm*) (*hurricane*)].

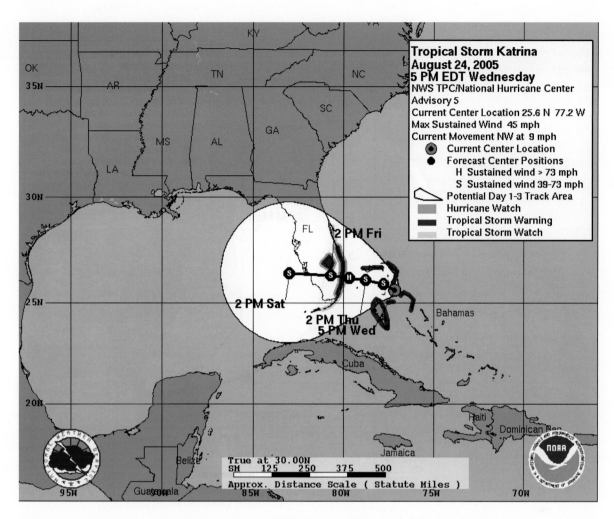

Figure 2. Advisory 5; 5 PM EDT, Wednesday, 24 August 2005.

Table 2 shows the Saffir-Simpson Hurricane Wind Scale. The category of hurricane strength is related to the damage potential and the measured sustained one-minute wind speeds in miles per hour (mph).

Table 2. Saffir-Simpson Hurricane Wind Scale

Category	Damage Potential	Sustained Winds (mph)
1	*Some*	74 - 95
2	*Extensive*	96 - 110
3	*Devastating*	111 - 130
4	*Catastrophic*	131 - 155
5	*Catastrophic*	greater than 155

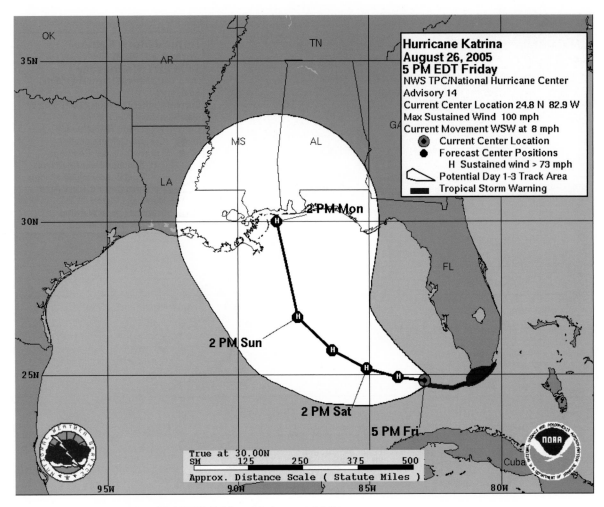

Figure 3. Advisory 14; 5 PM EDT, Friday, 26 August 2005.

15. At the Figure 3 advisory time, Katrina was a category **[(_1_)(_2_)(_3_)(_4_)(_5_)]** hurricane.

16. The Figure 3 forecast track issued at 5 PM EDT Friday, 26 August, projected Katrina to make landfall along the Gulf of Mexico coast near the Mississippi-Alabama border during the afternoon on **[(_Sunday_)(_Monday_)(_Tuesday_)]**.

17. Consider on Figure 3 the two locations: Mobile Bay, AL (coastal indentation to the right of the final plotted H dot) and New Orleans, LA (to the left of predicted landfall position and south of the irregular Lake Ponchatrain oval). Based on the forecast track, one would expect the higher storm surge from the major landfalling hurricane to occur in the **[(_Mobile_)(_New Orleans_)]** area.

Hurricane Katrina continued intensifying to a category 5 storm with winds of 173 mph before finally weakening to a strong category 3 at landfall near the Louisiana-Mississippi border. Actual landfall was to the west of the landfall location predicted on the Figure 3 image. The massive size of the storm caused devastating storm surges from Louisiana to the Florida panhandle with a maximum surge of 27.8 feet at Pass Christian, MS. Ironically, category 5 Hurricane Camille in 1969 had a maximum surge of about 32 feet also near Pass Christian.

Of course, a major part of the Katrina story was the death and devastation in New Orleans. Due to Katrina's large size, winds prior to landfall caused water from Lake Pontchartrain to slosh over seawalls and canal banks, undermining and cutting levees, and flooding the city. About 80% of the city flooded, in some places up to 20 feet deep. The last of the water was not removed until 43 days later. The direct and indirect death toll from Katrina was estimated near 2000, mostly in New Orleans. The total cost was estimated at $81 billion, making Katrina the nation's most costly hurricane ever.

The NHC Katrina advisories website provides an animation of the forecast graphics for all advisories issued: (*http://www.nhc.noaa.gov/archive/2005/KATRINA_graphics.shtml*). The user may adjust the speed or even stop the sequence and zoom in on a particular graph for study.

As directed by your course instructor, complete this investigation by either:

1. *Going to the Current Weather Studies link on the course website, or*
2. *Continuing to the Applications section for this investigation that immediately follows in this Investigations Manual.*

Investigation 12A: Applications

HURRICANES

The 2009 Hurricane Season in the Atlantic Basin was relatively tame compared to the 2008 season and several other recent years. (The Atlantic Basin Hurricane Season officially runs from June 1st to November 30th. The eastern North Pacific Basin season extends from May 15th to November 30th.) A descriptive summary of the 2009 Atlantic Basin season is given at: *http://www.noaanews.noaa.gov/stories2009/20091130_endhurricaneseason.html*. The 2009 season in the Atlantic basin had only three hurricanes (Bill, Fred, and Ida) and six tropical storms with a mere two tropical storms making U.S. landfalls (one being the weakened Hurricane Ida). Averages for the Atlantic basin are 10 named storms, 6 becoming hurricanes and 2 being major hurricanes - based on occurrences from 1950 to 2000. Hurricanes Bill and Fred did achieve major hurricane status, Saffir-Simpson category 3 or higher.

Although the 2009 season was relatively quiet, other recent seasons have suggested a trend to above average numbers of tropical storms. There has also been speculation that global warming of the ocean surface waters may spawn more hurricanes. Studies seem to suggest that there may be no more hurricanes than average (a number with great variability!) but that they may be somewhat stronger in a warmer climate. The weak El Niño conditions experienced in early 2010 were expected to become more neutral by the onset of the 2010 hurricane season, and possibly transition to La Niña conditions during the summer, suggesting above average Atlantic tropical activity in 2010 (based on May 2010 predictions). Of course, it is uncertain whether the storms that do form will remain at sea or make landfall.

Figure 4 is the map of the tracks of Atlantic Basin tropical weather systems for the 2009 season from NOAA's National Hurricane/Tropical Prediction Center. The track of each named storm is identified by the number at the beginning and the end of its track. Solid dots denote the location of the tropical system's center at 00Z and open circles with the date at its 12Z position. The storm's strength (whether tropical depression, tropical storm, hurricane or major hurricane) is color coded according to the scale at the lower right of the map area. Recall that hurricane threshold is 64 kts (74 mph). The original map is found at *http://www.nhc.noaa.gov/2009atlan.shtml*.

18. The majority of Atlantic tropical systems travel westward at lower latitudes and then recurve to the north and northeast upon reaching the belt of Westerly winds. With this directional pattern, storms that <u>do not</u> recurve until reaching about 75° W [(***will***)(***will not***)] likely make a landfall before recurvature northward.

19. An important function of hurricanes and tropical cyclones in the Earth-atmosphere system is the transport of heat and moisture from tropical oceans to land or ocean surfaces in the higher latitudes. From the 2009 path of Hurricane Bill (#2) which originated over the tropical Atlantic Ocean, the final position implies that this disturbance [(***would***)(***would not***)] transport energy and moisture to higher latitudes. Other Atlantic tropical cyclones of 2009 did not exist long enough to follow this general pattern.

20. Hurricane Bill achieved major Saffir-Simpson Category 3 strength during its lifetime. According to Table 2 in this investigation, sustained wind speeds would have been [(*74 - 95*)(*96 - 110*)(*111 - 130*)(*131 - 155*)(*over 155*)] mph.

21. Because of the storm's counterclockwise circulation around its low-pressure center, such winds would have caused extensive wind damage, accompanying heavy rainfall and a damaging storm surge mainly to the [(*left*)(*right*)] of the point where an advancing hurricane's center came ashore. Bill was only a tropical storm when it finally crossed Newfoundland. Nevertheless, a swimmer drowned in Florida and one person was swept into the surf in Maine due to Bill's wave action. Also, Tropical Storms Claudette and Ida did make U.S. landfalls with their associated damage.

22. **Figure 5** (*http://www.solar.ifa.hawaii.edu/Tropical/GifArchive/wld2009.gif*) is a map from the University of Hawaii of the tracks of tropical weather systems worldwide for the 2009 season. Storm maximum wind speed, implying strength, is shown by color coding with an arrowhead indicating direction of motion. A horizontal line has been added across the map denoting the position of the equator. Tropical storms of 2009 [(*did*)(*did not*)] cross the equator from one hemisphere to the other.

23. This lack of tropical activity at the equator (0° latitude) [(*was*)(*was not*)] due to the absence of the Coriolis Effect at the equator providing rotation for storm formation.

24. From Figure 5 one can note that the greatest tropical cyclone activity in 2009 occurred in the [(*North Atlantic*)(*western North Pacific*)(*Southern Indian*)(*South Atlantic*)(*South Pacific*)] ocean basin. This frequency of storms is typical and caused by the large expenses of warm ocean waters persisting year round in that part of the world. (You may recall the sea-surface temperatures plotted in Investigation 9B.)

25. **Figure 6** (*http://www.solar.ifa.hawaii.edu/Tropical/GifArchive/wld2010.gif*) from the University of Hawaii of tropical cyclones paths for 2010 up until 21 May showed that tropical cyclones [(*had*)(*had not*)] already occurred, mainly in the Southern Hemisphere where the seasons are reversed and waters retain much of their summer warmth.

26. Also a tropical depression and a tropical storm had formed in the western North Pacific. From those occurrences one can surmise that sea-surface temperatures and upper atmospheric wind patterns [(*were*)(*were not*)] favorable for formation of tropical cyclones in the western North Pacific by early May.

For information on tropical cyclones anywhere in the world, any time of the year, check the University of Hawaii web site: *http://www.solar.ifa.hawaii.edu/Tropical/tropical.html*. For information on Atlantic storms, go to the National Hurricane Center's webpage, *http://www. nhc.noaa.gov/*. On the menu under **Hurricane History**, *Seasons Archive*, track maps and individual storm reports for current and past years can be found. Another valuable item under **Hurricane Awareness** is a list of *Frequent Questions*.

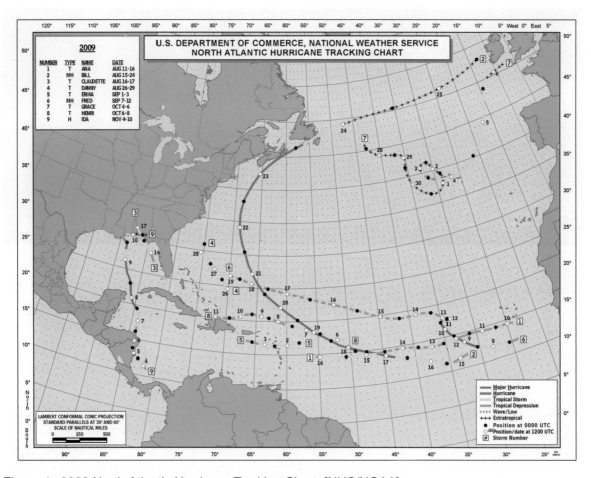

Figure 4. 2009 North Atlantic Hurricane Tracking Chart [NHC/NOAA]

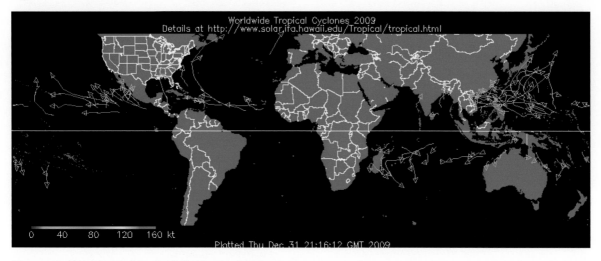

Figure 5. Worldwide Tropical Cyclones 2009

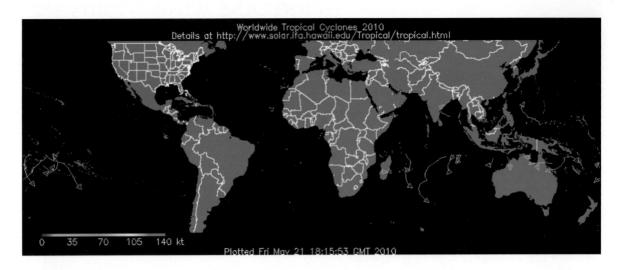

Figure 6.
Worldwide Tropical Cyclones, 1 January 2010 to 21 May 2010.

HURRICANE WIND SPEEDS AND PRESSURE CHANGES

Objectives:

Hurricanes are intense tropical cyclones spawned and sustained over warm ocean waters. They are fueled primarily by the energy transported to the atmosphere from the ocean through evaporation and subsequent condensation of water vapor within the hurricane eyewall and spiral-band cumuliform clouds. Latent heat is released as water vapor condenses, warming air that expands and rises as more humid air flows upward from near the ocean surface to replace it. Within the storm's center or eye, air sinks and warms adiabatically. The less dense air in this warm eye exerts a lower pressure than the surrounding atmosphere, producing intense inward-directed horizontal pressure gradients that drive the air motions which result in hurricane-force winds. Low central pressures, high eyewall wind speeds, and heavy rains continue unabated until the energy supply to the hurricane is disrupted. Weakening of the system could result from travel over colder ocean waters, which reduces evaporation, or an encounter with land which limits the supply of water vapor while its surface roughness (friction) slows the winds and brings chaos to its organized structure.

After completing this investigation, you should be able to:

- Describe the relationship between the maximum wind speeds and the central pressure in a hurricane.
- Categorize the damage potential of a hurricane based on wind speeds.
- Explain how wind speeds in hurricanes are affected by landfall.

Wilma:

1. **Table 1** gives the position, central pressure, and maximum sustained wind speed of the tropical cyclone that evolved into Hurricane Wilma during 17 – 25 October 2005 at 6-hourly intervals. The central pressure of the system at 00, 06, 12, and 18 UTC is graphed in **Figure 1**, with the pressure scale to the left. From Table 1, at what October 2005 date and time (date/UTC or Z time) was the central pressure lowest? [(**_19/1200_**) (**_21/1800_**)(**_24/1200_**)]

2. At this time the central pressure was [(**_950_**)(**_926_**)(**_882_**)] mb. This most intense condition occurred as Wilma was slowly crossing the warm waters of the Caribbean Sea east of the Yucatán Peninsula.

3. From 18/18 to 19/12 the total pressure fall was [(**_72_**)(**_93_**)(**_117_**)] mb.

4. This pressure fall occurred over a period of 18 hours, decreasing at the rate of about [(**_4.0_**)(**_5.2_**)(**_6.5_**)] mb/hr.

Table 1. Best track for Hurricane Wilma, 17–25 October 2005. From: National Hurricane Center, (*http://www.nhc.noaa.gov/pdf/TCR-AL252005_Wilma.pdf*).

Date/time	Lat. (°N)	Long. (°W)	Pressure (mb)	Wind speed (kts)
17/1200	16.3	79.7	999	40
17/1800	16.0	79.8	997	45
18/0000	15.8	79.9	988	55
18/0600	15.7	79.9	982	60
18/1200	16.2	80.3	979	65
18/1800	16.6	81.1	975	75
19/0000	16.6	81.8	946	130
19/0600	17.0	82.2	892	150
19/1200	17.3	82.8	882	160
19/1800	17.4	83.4	892	140
20/0000	17.9	84.0	892	135
20/0600	18.1	84.7	901	130
20/1200	18.3	85.2	910	130
20/1800	18.6	85.5	917	130
21/0000	19.1	85.8	924	130
21/0600	19.5	86.1	930	130
21/1200	20.1	86.4	929	125
21/1800	20.3	86.7	926	120
22/0000	20.6	86.8	930	120
22/0600	20.8	87.0	935	110
22/1200	21.0	87.1	947	100
22/1800	21.3	87.1	958	85
23/0000	21.6	87.0	960	85
23/0600	21.8	86.8	962	85
23/1200	22.4	86.1	961	85
23/1800	23.1	85.4	963	90
24/0000	24.0	84.3	958	95
24/0600	25.0	83.1	953	110
24/1200	26.2	81.0	950	95
24/1800	28.0	78.8	955	105
25/0000	30.1	76.0	955	110
25/0600	33.3	72.0	963	100
25/1200	36.8	67.9	970	90
25/1800	40.5	63.5	976	75

5. The most rapid "deepening" or lowering of Wilma's central surface air pressure resulting in storm intensification was [(*4*)(*7*)(*9*)(*11*)] mb/hr from 19/00 to 19/06.

Using the wind speed values from Table 1, plot on the Figure 1 graph the wind speeds every 6 hours. Use the wind speed scale along the <u>right</u> side of the graph. Plot each value with a dot. When completed, connect adjacent dots with dashed straight lines to make a continuous wind speed time series. The first two dots, at 17/12 and 17/18, have already been plotted for you and connected with a straight line.

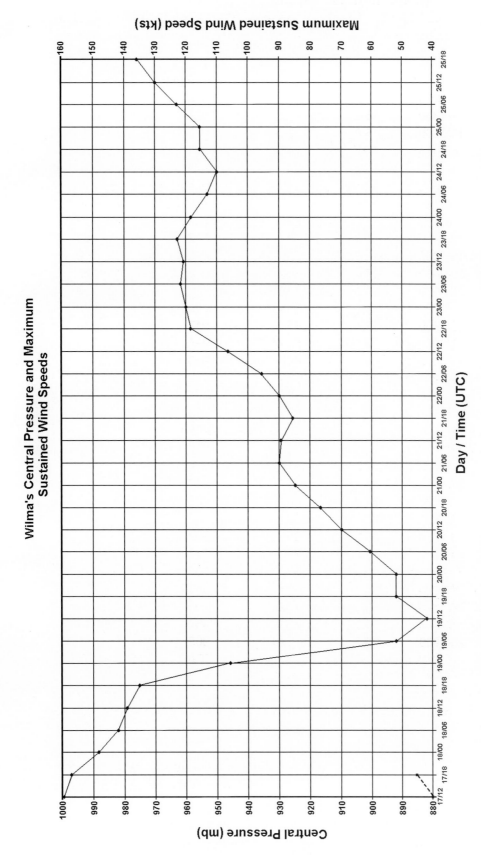

Figure 1. Graph of Wilma's central pressure (solid curve) and maximum sustained wind speeds (dashed curve) from 17 - 25

6. From your completed graph, at what date and time was the wind speed greatest? [(*19/12*) (*24/06*)(*25/00*)]

7. This was [(*an earlier*)(*the same*)(*a later*)] time compared to the time of lowest central pressure.

8. Wilma's center crossed the tip of the Yucatán Peninsula in Mexico from about 00 UTC on 22 October to 00 UTC on 23 October. Place a bracket along the time axis at the bottom of the Figure 1 graph to denote these times and label it with "YP". This overland time was for the center of the storm, whereas extreme weather conditions associated with the system extended from its eye wall outward several tens of kilometers in all directions. During the initial several hours of the overland time period, the wind speeds generally [(*increased*)(*decreased*)] as the storm's circulation responded to the increased roughness of the underlying land surface.

9. When its maximum wind speeds are greater than 64 kts (74 mph), a tropical cyclone is classified as a hurricane. Wilma moved rapidly northeastward from the Yucatán Peninsula before crossing the coast of southern Florida at 24/1030 (1030 UTC on 24 October). According to the graphed data, was Wilma at hurricane strength when it made landfall in Florida? [(*Yes*)(*No*)].

As directed by your course instructor, complete this investigation by either:

1. *Going to the Current Weather Studies link on the course website, or*
2. *Continuing to the Applications section for this investigation that immediately follows in this Investigations Manual.*

Investigation 12B: Applications

HURRICANE WIND SPEEDS AND PRESSURE CHANGES

The first part of Investigation 12A dealt with the path of Hurricane Katrina and its damaging storm surge. In this investigation, we consider Hurricane Wilma's landfall just south of Naples, FL. Hurricane Wilma is notable, in that, while still over the Gulf of Mexico it had the lowest central sea-level air pressure ever measured in an Atlantic basin hurricane.

The graph analyzed and interpreted in the initial part of this investigation clearly showed how the maximum sustained wind speed was related to the central pressure of the hurricane. We saw from earlier investigations that the horizontal pressure gradient force is the principal control of wind speed. The low pressure in the hurricane eye compared to the near-normal sea-level pressures surrounding the storm produces a very strong horizontal air pressure gradient and, hence, the high wind speeds.

Figure 2 is a plot of the track of Wilma's center of circulation along with maximum sustained wind speeds at various times denoted by the squares. (You are strongly urged to view an animation of the track shown by the NWS Miami radar by going to *http://www.srh.noaa.gov/mfl/?n=wilma*, and scrolling down to Figure 3. There, click on the low resolution link.)

10. **Center a coin such as a quarter on the Figure 2 printed track in the Gulf of Mexico to represent Hurricane Wilma. As you move the coin northeastward along the hurricane's track, also rotate it counterclockwise to simulate the surface wind directions of the traveling hurricane.** (The motions are not to scale; see the radar animation noted above.) With your rotating coin/hurricane approaching the Florida coast, the wind direction at Naples, FL, is generally <u>toward</u> the **[(*southeast*)(*northwest*)]**.

11. After your coin/hurricane passes Naples and is over the Florida peninsula, the wind direction at Naples blows generally <u>toward</u> the **[(*southeast*)(*northwest*)]**.

Figure 3 is a plot of the water levels, winds and air pressures at Naples as measured by instruments associated with the NOAA tidal station at Naples harbor. In the middle panel of Figure 3, wind speeds and directions are presented. The time period across the base of the graph is from 9:00 am EDT 23 October 2005 to 9:00 am EDT 26 October 2005. The vertical axis denotes wind speed in knots. The time of the last plotted observation is noted on each panel just preceding the dashed vertical line. Wind speeds are denoted by red dots with directions shown using attached blue arrows. The center of the large circulation of Wilma officially made landfall at 7 a.m. on 24 October a few miles south of Naples at Everglades City. **After carefully determining the 24 October 7 a.m. tick on the time scale, draw a vertical line across both the middle wind and lower pressure panels at this landfall time.**

12. For several hours prior to landfall, the wind direction at Naples exhibited a wind component generally <u>toward</u> the **[(*east*)(*west*)]**

13. This direction of air motion **[(*was*)(*was not*)]** generally consistent with your coin/ hurricane circulation model.

14. The most dramatic shift in wind direction took place within a short time of the nearby landfall of Hurricane Wilma. In the twelve or so hours following the dramatic wind shift, the wind directions at Naples were generally <u>toward</u> the **[(*southeast*)(*west or northwest*)]**.

15. The *maximum* wind speed at Naples, shown by the red dots in the middle panel (be sure to note the dots among the text data in the same panel), was about **[(*74*)(*81*)(*85*)]** knots (approx. 85 mph) at 8:00 a.m. on 24 October.

16. The *minimum* air pressure (lower panel) was about **[(*962*)(*974*)(*981*)]** millibars which occurred an hour earlier.

17. Comparing the middle and lower panels of Figure 3, it can be seen that in general, as air pressures were <u>decreasing</u>, wind speeds were **[(*decreasing*)(*increasing*)]**.

18. And as air pressures began <u>increasing</u>, **[(*wind speeds decreased*)(*wind directions shifted dramatically*)]**.

19. The top panel of Figure 3 displays the actual water levels above the mean lower low water (MLLW) reference level by a series of red crosses. The predicted tide level (which does not include weather impacts) is depicted by the smooth blue curve. The highest actual water level shown was about **[(*1*)(*3*)(*5*)]** foot (feet). This occurred at about 12 noon on 24 October.

20. As shown by the smooth blue curve, this was also the approximate time of predicted **[(*low*)(*high*)]** tide. [High tides occur at the crests of the blue predicted tide level curve; low tides occur at the troughs of the curve.]

21. At its peak, the storm surge was **[(*1.5*)(*3.5*)(*4.5*)]** feet higher than the height of the predicted tide. Had this storm surge occurred near high tide several hours earlier or later, the surge would have been much more damaging. This contrasts with Katrina, which was particularly devastating because the highest surge occurred near the time of high tide.

For more information on Hurricane Wilma, consult the Miami NWS website noted above, or that from the Key West NWS office (*http://www.srh.noaa.gov/key/?n=wilma* , or Tampa Bay NWS office (*http://www.srh.noaa.gov/tbw/?n=tampabayweatherhurricanewilma*). The National Hurricane Center website (*http://www.nhc.noaa.gov/*) also contains a summary of the season as well as studies of individual storms.

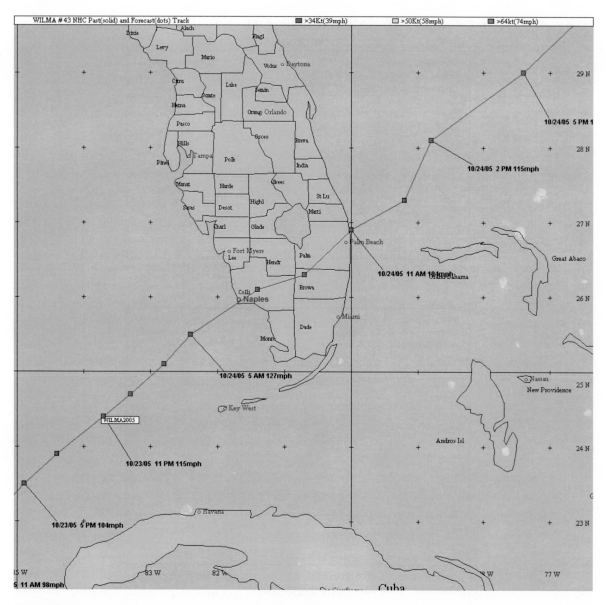

Figure 2.
The track of the center of circulation of Wilma as it neared Florida. [NOAA/NWS Weather Forecast Office – Miami]

Suggestions for further activities: Explore other information provided by NOAA's National Hurricane Center's website, particularly if you or your relatives live or visit along the Gulf or East Coasts, areas prone to hurricane landfall. For information on tropical storms anywhere in the world at any time of the year, check the University of Hawaii website: *http://www. solar.ifa.hawaii.edu/Tropical/tropical.html*.

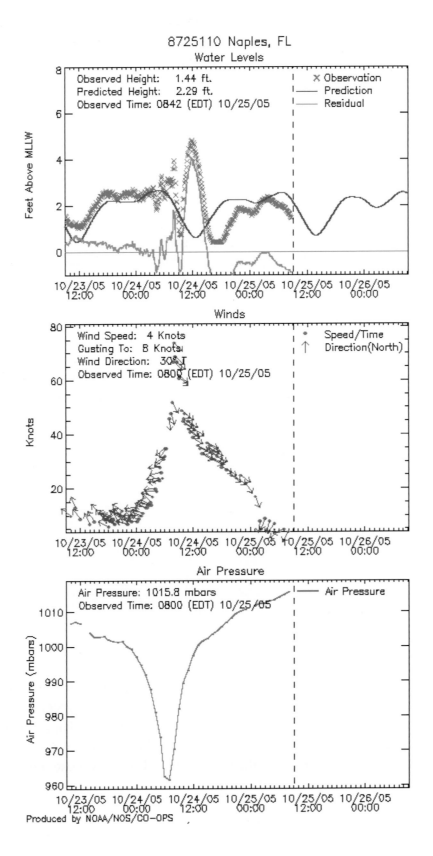

Figure 3.
Water levels, winds and air pressures from the NOAA tidal station at the Naples, FL harbor as Wilma made landfall.

WEATHER INSTRUMENTS AND OBSERVATIONS

Objectives:

Weather is the state of the atmosphere at a particular place at a given time. We must describe as closely as possible the state of this mixture of gases with minute quantities of particles (e.g., cloud droplets, ice crystals, aerosols) we call the atmosphere at the time and location where instruments are available. We are interested in knowing for initial descriptive purposes and ultimately for predictive reasons, the atmosphere's heat energy, density, large-scale motions, water vapor concentration, and liquid or solid water in clouds and precipitation. These quantities translate to the more common parameters of temperature, atmospheric pressure, wind speed and direction, dewpoint (or relative humidity), cloud cover (including height), visibility, precipitation (amount and type), and the general character of the "weather."

Weather observations are taken by international convention at three-hourly intervals each day, at 0000 UTC, 0300 UTC, etc. [Universal Time Coordinated (UTC), also known as Z time or Greenwich Mean Time (GMT), is the time along the prime meridian, 0 degree longitude.] For aviation purposes, weather observations are routinely reported every hour, twenty-four hours a day. In the U.S., most weather observations are taken by the Automated Surface Observing System (ASOS) sensors.

After completing this investigation, you should be able to:

- Describe the Automated Surface Observing System (ASOS) and the data it provides.
- Describe how to access weather observations for the U.S. and the world via the Internet.

Introduction:

Weather observations are a combination of direct measurements by various sensors and determinations made by computer algorithms (programs based on several values). For example, sensors determine the air temperature, dewpoint, pressure, ceiling, precipitation, and wind. The amount of sky cover is determined by the fraction of ceilometer beams that detect cloud or clear air. Visibility is computed from the amount of light scattered back to a sensor from a small volume of air surrounding the sensor. "Weather" type is a combination of temperature, dewpoint, and visibility values and may specify haze, smoke or fog.

The Automated Surface Observing System (ASOS) consists of an array of instruments including an electronic thermometer, an electronically chilled mirror and light or absorptive humidity sensor to determine the dewpoint, an anemometer and wind vane or sonic anemometer for wind speed and direction, a vertical pulsed laser ceilometer for cloud height and amount, pulsed laser instruments for detecting scattered light for visibility and weather type, a vibrating column for detecting freezing rain, and a heated tipping bucket or weighing rain gauge for rain and snowfall. (For complete technical details of ASOS sensors, see: *http://www.nws.noaa.gov/asos/.*)

The National Weather Service (NWS) provides an Internet site that allows you to obtain the latest weather observation from ASOS instruments at any particular location, selected by choosing the state and then the city at: *http://weather.noaa.gov/*. Times are given in UTC with equivalents in local times. Regardless of location, all times are given as Eastern Standard or Daylight, as appropriate with an initial conversion. A sample observation for Des Moines International Airport, Iowa, follows:

Current Weather Conditions: Des Moines International Airport, IA, United States
(KDSM) 41-32N 93-40W 295M

Conditions at Apr. 30, 2010 – 02:54 PM EDT   *2010.04.30 1854 UTC*

Wind	from the SSE (150 degrees) at 14 MPH (12 KT)
Visibility	10 mile(s)
Sky conditions	Overcast
Precipitation last hour	0.01 inches
Temperature	62.1 F (16.7 C)
Dew Point	54.0 F (12.2 C)
Relative Humidity	74%
Pressure (altimeter)	29.35 in. Hg (993 hPa)
Coded observation	KDSM 301854Z 15012KT 10SM BKN075 BKN130 OVC180 17/12 A2935 RMK AO2 RAE15 SLP933 P0001 T01670122

Notes: Times of observation, taken within 10 minutes of the top of the hour, are ordinarily considered to have been acquired at the top of the hour. The readings above were taken at 1854 UTC, but are considered to be the 1900 UTC readings. The pressure unit hectoPascals (hPa) is numerically equivalent to millibars (mb). Standard sea level pressure of 29.92 in. Hg (inches of mercury) = 1013.25 hPa = 1013.25 mb.

1. Additional information is also available within the coded observation. For example, the fourth, fifth and sixth letter-number groups contain information on the amount of sky covered by low, middle and high level clouds. Here, clouds were "broken" at 7500 ft., broken at 13000 ft. and overcast at 18000 ft. Consequently, the portion of the entire sky that was covered by clouds when <u>all levels</u> were collectively considered was **[(*scattered*) (*broken*)(*overcast*)]**.

Another item listed in the report following the current weather conditions is the maximum and minimum temperatures over recent 6- and 24-hour periods. If precipitation had occurred, total precipitation within certain time periods is also listed.

Maximum and Minimum Temperatures

Maximum F (C)	Minimum F (C)	
64.9 (18.3)	57.9 (14.4)	In the **6 hours** preceding Apr. 30, 2010 - 1:54 PM EDT / 2010.04.30 1754 UTC
82.0 (27.8)	57.0 (13.9)	In the **24 hours** preceding Apr. 30, 2010 – 1:54 AM EDT / 2010.04.30 0554 UTC

2. The range of temperature (difference between maximum and minimum) over the 24 hours preceding the 0554 UTC 30 Apr 2010 observation time was [(*12.1*)(*13.9*)(*25.0*)] F°.

Following the maximum and minimum temperatures and precipitation report (if any) for the particular hour is an hour-by-hour listing (time series) of the weather conditions for the preceding twenty-four hours. In this case:

Date	Time EDT (UTC)	Temperature F (C)	Dew Point F (C)	Pressure Inches (hPa)	Wind MPH	Weather
Apr 30	3 PM (19)	62.1 (16.7)	54.0 (12.2)	29.35 (993)	SSE 14	
	2 PM (18)	60 (16)	55 (13)	29.34 (993)	SSE 8	light rain
	1 PM (17)	57 (14)	53 (12)	29.36 (994)	WSW 5	light rain, mist
	Noon (16)	59 (15)	53 (12)	29.38 (994)	W 10	light rain
	11AM (15)	60 (16)	55 (13)	29.32 (992)	W 6	light rain
	10 AM (14)	64.0 (17.8)	59.0 (15.0)	29.32 (992)	W 9	
	9 AM (13)	64 (18)	59 (15)	29.31 (992)	WSW 7	
	8 AM (12)	64 (18)	57 (14)	29.28 (991)	SW 6	
	7 AM (11)	63.0 (17.2)	57.9 (14.4)	29.21 (989)	SE 15	light rain
	6 AM (10)	63.0 (17.2)	57.0 (13.9)	29.27 (991)	ESE 12	light rain
	5 AM (9)	62 (17)	57(14)	29.3 (992)	ESE 8	light rain with thunder
	4 AM (8)	62 (17)	57 (14)	29.28 (991)	ESE 8	light rain with thunder
	3 AM (7)	69.1 (20.6)	54.0 (12.2)	29.28 (991)	SE 10	
	2 AM (6)	71.1 (21.7)	54.0 (12.2)	29.27 (991)	SE 9	
	1 AM (5)	71.1 (21.7)	55.9 (13.3)	29.27 (991)	S 12	
	Midnight (4)	72.0 (22.2)	55.9 (13.3)	29.27 (991)	S 13	
Apr 29	11 PM (3)	73.0 (22.8)	55.9 (13.3)	29.27 (991)	SSE 12	
	10 PM (2)	73.9 (23.3)	55.9 (13.3)	29.26 (990)	SSE 15	
	9 PM (1)	75.9 (24.4)	57.0 (13.9)	29.25 (990)	S 14	
	8 PM (0)	78.1 (25.6)	55.9 (13.3)	29.26 (990)	S 21	
	7 PM (23)	81.0 (27.2)	57.0 (13.9)	29.26 (990)	S 20	
	6 PM (22)	81.0 (27.2)	57.0 (13.9)	29.26 (990)	S 24	
	5 PM (21)	82.0 (27.8)	55.9 (13.3)	29.26 (990)	S 21	
	4 PM (20)	81.0 (27.2)	57.0 (13.9)	29.23 (989)	S 22	

Use the table of hourly observations above to answer the following four questions:

3. The highest temperature during the period was [(*64.8*)(*82.0*)(*83.4*)] °F.

4. The lowest dewpoint during the period was [(*51*)(*53*)(*59*)] °F.

5. During the period being reported, the minimum pressure was [(*29.21*)(*29.26*)(*29.31*)] in.

6. The weather condition of light rain was reported in [(*two*)(*six*)(*eight*)] of the hourly observations.

The following is from the AMS course website and is a partial listing of the State Surface Data - Text for IA (Iowa) at 19Z (1900 UTC), essentially the same time as the NWS observation given above.

Data for IA

19Z 30 APR 2010

STN	TMP	DEW	DIR	SPD	GST	CLDL	CLDM	CLDH	ALT	PMSL	PTD	WTHR	PCPN
CID	63	62	15	8		BKN			29.32	992.3			
BRL	74	62	17	13	19	OVC			29.38	994.6			
DSM	**62**	**54**	**15**	**12**		**BKN**	**BKN**	**OVC**	**29.35**	**993.3**			
OTM	64	63	25	10		OVC			29.38	994.3		R	
DBQ	71	59	19	18	25	BKN			29.38	994.4			
ALO	63	60	25	13		OVC			29.31	992.0		R-	
MCW	63	58	19	8		SCT			29.28	990.7			
SUX	66	48	24	15		SCT			29.28	991.2			
EST	59	52	25	13		BKN			29.27	991.1			
DVN	74	59	19	20	28	BKN			29.41	995.5			
IOW	66	65	15	7		BKN			29.37	994.2		R	
SPW	61	48	24	17		BKN			29.29	991.5			
MIW	60	57	17	8		OVC			29.31	992.2		R-	
AMW	62	57	16	10			SCT		29.32	992.2			
LWD	61	55	16	6		CLR			29.38	994.4			

7. In this text listing for IA, the temperature (TMP) and dewpoint (DEW) are in whole degrees Fahrenheit, wind: direction (DIR) in tens of degrees, speed (SPD) and gusts (GST) in knots, and sea level pressure (PMSL) in tenths of hectoPascals. The "sky cover" or amount of cloudiness is reported as clear (CLR) – no clouds, scattered (SCT) – less than half covered, broken (BKN) – more than half covered, or overcast (OVC) – completely cloudy for each of the low (CLDL), middle (CLDM) and high (CLDH) levels. Precipitation (PCPN) is the amount in hundredths of inches over the last three hours. Compare the NWS report for Des Moines International Airport in item #1 (KDSM) with the report for the same time for Des Moines International Airport (DSM) from this question's text listing for IA. Note that the two reports do not report all the same weather conditions. One condition not listed both reports is [(*wind speed*)(*relative humidity*)(*dewpoint*)].

8. The course website delivers national and regional maps on which surface weather data are plotted. The data on these maps are updated hourly. **Figure 1** is a sample national map (Isobars, Fronts, Radar & Data) which displays data collected at 19Z (3 pm

Eastern Daylight Time, 2 pm CDT, etc.) on 30 APR 2010. (The station identifiers for the reporting stations can be found by calling up the "Available Surface Stations" map on the course website. The identifiers shown on that map do not include a "K" that is the first letter of all contiguous United States station identifications.) Surface weather data are plotted on the national map in, on, and around a circle representing the station. Temperature, in Fahrenheit degrees, is plotted at the "11 o'clock" position relative to the station circle. The temperature at Indianapolis, Indiana, at Figure 1 map time was [(*60*)(*75*)(*82*)] °F.

9. The national map displays surface observational data from a sufficient number of stations to determine large-scale weather patterns and features. Temperatures reported on the Figure 1 map show that the area of the nation where the highest temperatures prevail was over the [(*northern states*)(*Southwest*)(*Southeast*)].

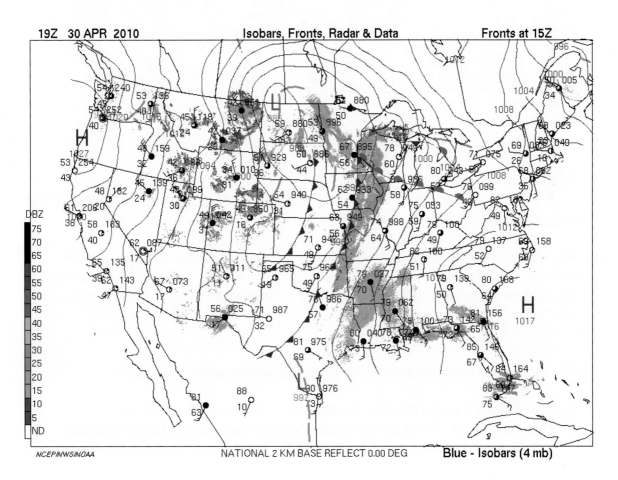

Figure 1. U.S. – Surface data map for 19Z 30 APR 2010.

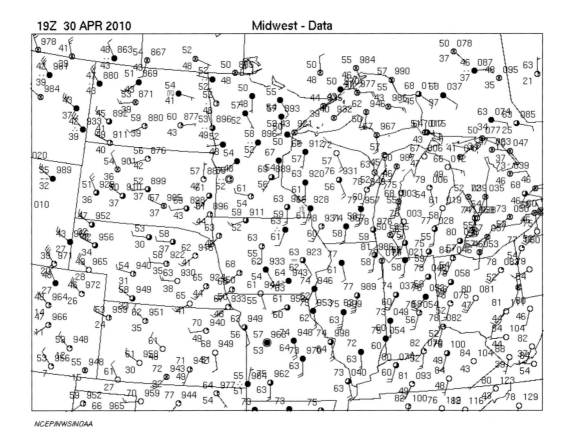

NCEP/NWS/NOAA

Figure 2. Midwest – Surface data map for 19Z 30 APR 2010.

10. **Figure 2** is a sample regional map, one of nine including Alaska and Hawaii, provided via the course website. This map is labeled [(***Southern Plains - Data***)(***Southeast - Data***) (***Midwest – Data***)]. This map is for the same time as the national map, Figure 1.

11. The regional maps display many more stations, allowing for more detailed weather analysis. For a comparison of the station densities on the two maps, the national map has two stations plotted in South Dakota whereas the number of stations on the regional map is [(*7*)(*10*)(*14*)].

As directed by your course instructor, complete this investigation by either:

1. *Going to the Current Weather Studies link on the course website, or*
2. *Continuing to the Applications section for this investigation that immediately follows in this Investigations Manual.*

Investigation 13A: Applications

WEATHER INSTRUMENTS AND OBSERVATIONS

From the course website, under the **Surface** section, click on "Meteograms for Selected Cities". From the meteograms page, select your nearest city from the map or table listing. Print that page.

12. The meteogram is constructed to portray [(*6*)(*12*)(*25*)] hourly observations over the time period shown.

Go to *http://www.weather.gov/view/national.php?map=on*. [This may be found from the course website by clicking on "Additional Surface Links" under the Surface section, and then, under United States Weather and Overview, click on 'NWS "Weather Page"'.] Once at the National Weather Service website, from the list to the left under Observations, click on "Surface Weather…" There, under Automated Surface Stations (ASOS), click on the "Select by State" drop box and choose a state. Under Current Weather Conditions, select a location in that state and click on "Go". Finally, scroll down to the 24 Hour Summary.

Compare the course website's meteogram with the NWS "Weather Page" data for the same station. Times are given in UTC along the horizontal axis labeled at the bottom of the meteogram increasing from left to right. Time for the tabular 24-hour summary observations from the NWS website is in Eastern Time (or selectable local time) with the UTC in parenthesis. Note that the listing starts with the most recent at the top. Choose several hours that are common to both the meteogram plot and the NWS data table. Draw vertical lines across the meteogram at those hours. Compare the temperature, dewpoint and wind speed and wind direction for those common UTC hours as shown on the meteogram to those reported in the table.

13. The observation values for the same time as shown on the meteogram and listed in the table are [(*equivalent*)(*quite different*)].

Current weather observations can be found on the Internet in a variety of formats from the course website, National Weather Service websites, and others. Investigate these sources and their displays.

Suggestions for further activities:

For those in the contiguous U.S. - Regional maps are made available on the course website so that you can track changes in weather and weather features, such as cold fronts, as they progress across the country. Call up the regional map that best fits your location. Mark your present location on the regional map and on the national map (Figure 1). You can check the three-letter identifier of the nearest station to your location by selecting the "Available Surface Stations" map on the website – click on the map for a full listing of stations.

For those in Alaska, Hawaii/Pacific or Puerto Rico/Caribbean areas - Regional maps for the coterminous U.S. are available to provide a detailed track of changeable weather and weather features. For areas outside of the "Lower 48" on the course website, scroll down to the area with Alaska/Hawaii and Pacific/Puerto Rico and Caribbean links and click on your region's link. For Alaska and Hawaii - Eastern Pacific, select "Surface map: state" under your region's heading and mark your location on this regional map. For Puerto Rico - Caribbean, select "Surface: area observations," then mark your present location on this regional map and click on the red dot closest to your location for station information.

Objectives:

Modern weather forecasting is based on a process termed *numerical weather prediction* (NWP). In a computer (or numerical) model of the atmosphere the physical laws that govern fluid motions (Newton's laws of motion, the first law of thermodynamics, conservation of mass, etc.) are put into computational forms for computer calculation. Supercomputers distribute an initial set of data throughout a three-dimensional grid that represents the atmosphere from the surface to the upper stratosphere. The physical relationships between variables (*e.g.* temperature, pressure, winds, water vapor) are used to take the initial values of these quantities a short step forward in time. Those new values are then used to step ahead once more at all the grid points. This process is repeated in leap-frog fashion into the future.

The starting point for these calculations is weather observations. By international convention, weather observations are taken at 0000Z and 1200Z at both the surface and in the upper atmosphere, worldwide. These observational data are exchanged and collected at national weather centers. The U.S. National Centers for Environmental Prediction, National Oceanic and Atmospheric Administration (NCEP/NWS/NOAA), collect these data and input them to the computer forecast process. After running the computer program to simulate times into the future of 6, 12, 18, 24 hours, or longer, the predicted values throughout the atmosphere are related to the surface temperatures, dewpoints, cloud cover, wind, and precipitation probabilities that are familiar components of weathercasts. This computer-generated information is distributed to local NOAA/NWS Forecast Offices.

The final step is taken by NWS meteorologists at local forecast offices. They use their experience and knowledge of local influences on the weather to adjust the computer information to the final forecasts that are provided to the public. Also, private sector providers utilize the NWS weather data and products to produce value-added products for the special needs of their customers.

After completing this investigation, you should be able to:

- Describe the general elements of a weather forecast.
- Compare the forecasts available to the public by NWS forecast offices with resulting weather conditions.

Making Forecasts:

Forecasts of weather conditions are provided for the U.S. via maps and descriptive summaries similar to those available from the course website (Forecasts). Forecast maps are provided for various parameters and in varying degrees of detail out to the following 16 days beginning with surface and upper-air observations at either 0000 UTC (00Z) or 1200 UTC

(12Z). Forecast maps give predicted positions of surface weather features (high- and low-pressure centers and fronts) along with areas and types of expected precipitation for a certain future (*valid*) time, specified in the lower left margin of each map. These maps show the conditions at 6-hour intervals during the period or by animation for the entire period.

Figure 1 is the surface weather map for 12Z 30 APR 2010 (8 AM EDT). This is a broad depiction of the actual weather conditions across the country when a forecast cycle was initiated. These and other surface observed values along with upper air conditions and satellite data were distributed across a three-dimensional grid on an imaginary Earth and stepped forward in time utilizing the physical laws governing fluid behavior.

1. At map time (and assuming the typical eastward motion of weather systems), national conditions showed that there was a [(***fair weather high-pressure system***)(***stormy weather low-pressure system***)] advancing generally toward the east and would impact the Detroit, Michigan area within the next couple of days.

2. Radar echo shadings on the Figure 1 weather map indicated that there [(***was***)(***was not***)] a likelihood of precipitation crossing the Michigan region in the next couple of days.

3. The present weather conditions at Detroit at map time were: temperature 57 °F, dewpoint 48 °F, wind from the south at about 5 knots, and the sky was partly cloudy. Based on the map conditions shown, the knowledge that weather systems generally move across the country from west to east, and fronts typically rotate counterclockwise (as seen from above) around advancing centers of low pressure, a generalized expectation for conditions in Detroit might be: [(***cool with rain and winds from the north likely for the next day or so followed by colder temperatures***)(***warm with variable cloudiness, possible thunderstorms, and winds from the south over the next day or two***)].

Forecast output of numerical weather prediction models consists of numerical values extrapolated in several-hour steps for days into the future. As the length of the forecast period (from the starting point) increases, small errors magnify and the detail in expected specific values of weather variables declines. Meteorologists at NWS forecast offices around the country use this computer guidance information along with knowledge of local conditions (e.g., topography, proximity to bodies of water, and other surrounding surface conditions) to tailor weather forecasts to the local region of responsibility. For example, one obvious effect is higher temperatures in urban areas compared to surrounding rural areas especially when synoptic-scale winds are light. NWS forecasts are, in turn, disseminated to newspapers, television and radio outlets, and Internet sites for announcement to the general public.

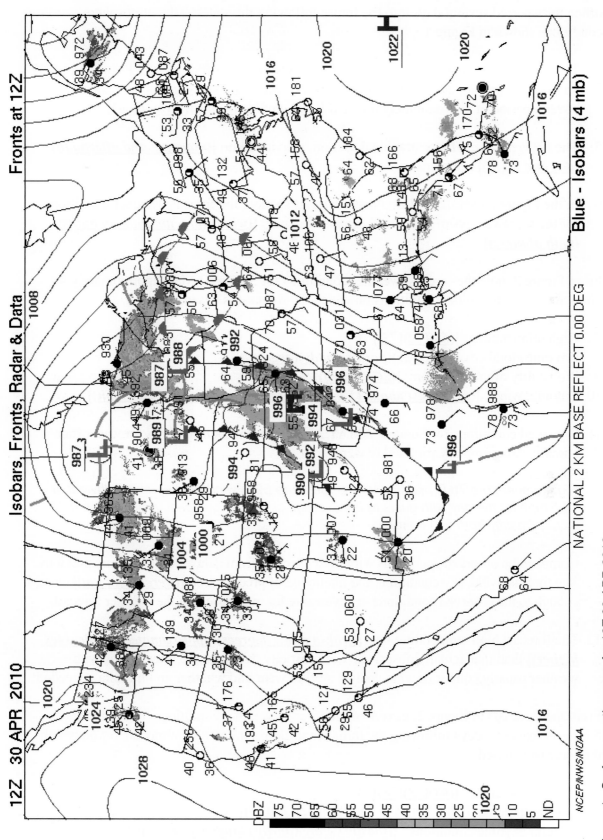

Figure 1. Surface weather map for 12Z 30 APR 2010.

Figure 2 is a partial display of the webpage forecast issued by the Detroit NWS forecast office for a period several days into the future following the state-of-the-atmosphere conditions shown in Figure 1.

4. The forecast was issued by the Detroit area NWS forecast office at 2:23 PM EDT (1823Z) on Friday, April 30, 2010. The *Forecast at a Glance* with small graphical depictions cover a period of approximately the next **[(*one day*)(*four days*)(*seven days*)]**.

5. The *Forecast at a Glance* graphical section is divided into **[(*morning and afternoon*) (*day and night*)]** segments.

6. In the Figure 2 *Detailed 7-Day Forecast* (text) section, the forecast weather conditions for the first two days of the period can also include: **[(*wind*) (*precipitation amounts*) (*both of these*)]**.

From Figure 2, note the forecast weather conditions for Sunday 2 May, approximately 48 hours following the initial conditions on which the forecast is based.

7. High temperature: **[(*68*)(*72*)(*75*)]** °F
8. Daytime cloud cover: **[(*clear*)(*partly cloudy*)(*overcast*)]**
9. Probability of precipitation: **[(*20*)(*50*)(*70*)]** %
10. Daytime wind: **[(*northwest about 15 mph*)(*south-southwest about 7-14 mph*)]**

11. Observed conditions for Sunday, 2 MAY 2010, from the Detroit NWS forecast office's climatic data summary for the day were:

    High temperature 76 F, morning low temperature 62 F, wind from south-southwest at 10.5 mph for the day with gusts to 28, average sky cover overcast, precipitation 0.77 inches (mainly from 6 to 8 am).

    Compare the observed conditions with those forecast from data at least 48 hours previous. The forecast wind, sky conditions and likelihood of precipitation **[(*were*)(*were not*)]** close enough to the observed conditions to have value for many users of the weather forecasts.

12. The forecast high temperature was **[(*within a few degrees*)(*approximately ten degrees higher*)]** than that actually observed. For most city dwellers, the interactions with weather usually consist of what coat (if any) to wear and whether an umbrella is needed!

From the forecasts in Figure 2, next note the forecast weather conditions for Wednesday 5 May, at least five days following the initial conditions (state of the atmosphere) on which the forecast is based.

13. High temperature: **[(*68*)(*73*)(*78*)]** °F
14. Low temperature (from Tuesday night): **[(*32*)(*37*)(*49*)]** °F
15. General daytime conditions: **[(*mostly sunny*)(*partly cloudy*)(*overcast*)]**

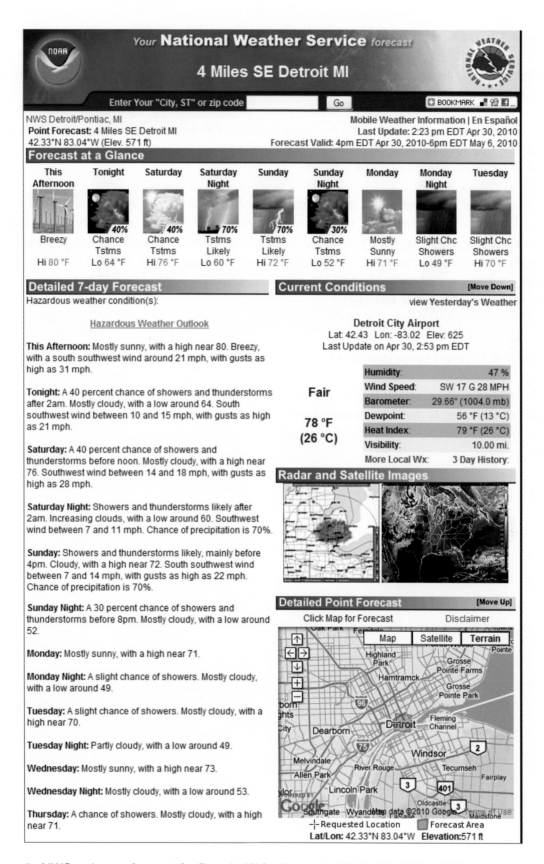

Figure 2. NWS webpage forecast for Detroit, MI for the period 30 April to 6 May 2010.

16. Mention of chance of showers for day or nighttimes: [(*yes*)(*no*)]

17. Observed conditions for Wednesday, 5 May 2010, from the NWS climatic data summary for the Detroit NWS forecast office:

High temperature 80 F, morning low temperature 54 F, wind from southwest at 11.6 mph for the day with gusts to 30, average sky cover partly cloudy, precipitation 0.16 inches (thunderstorm from 3 to 4 pm).

Compare the actual observed conditions for Wednesday with those forecast from the initial state of the atmosphere many days previous. The forecast temperatures and sky conditions [(*were*)(*were not*)] as close to the observations as those only a day or two from the initial forecast time. For general planning about a week into the future, these expectations would probably provide useful information beyond that of pure guessing for many users of weather forecasts.

As directed by your course instructor, complete this investigation by either:

1. *Going to the Current Weather Studies link on the course website, or*
2. *Continuing to the Applications section for this investigation that immediately follows in this Investigations Manual.*

Investigation 13B: Applications

WEATHER FORECASTS

"The National Weather Service (NWS) provides weather, hydrologic, and climate forecasts and warnings for the United States, its territories, adjacent waters and ocean areas, for the protection of life and property and the enhancement of the national economy. NWS data and products form a national information database and infrastructure which can be used by other governmental agencies, the private sector, the public, and the global community." (NOAA/ NWS Mission Statement)

The gathering of weather information is therefore a process that is conducted primarily so the NWS can fulfill its forecasting and warning mission. The Internet now forms an exceptional vehicle to disseminate those forecasts, warnings, and other weather and climate information to the public in text and graphic formats. The following Internet address: *http://www. weather.gov/* is the location for the latest official watches and warnings from the National Weather Service. The areas on this U.S. map that currently have active warnings, watches, advisories, and/or special weather statements are colored according to the legend below the map. **Figure 3** is the watches and warnings map for 30 April 2010 as an example.

18. Find your location on the National Weather Service interactive map and click on it. The resulting webpage shows the local NWS forecast office's (location at the top of the page) area of responsibility for issuing weather warnings and forecasts. From the regional map, click on your approximate location. Examine the forecast products that are displayed. Under the pictorial forecasts of the next few days are detailed forecasts. These detailed descriptive forecasts are presented out to [(*3*)(*5*)(*7*)] days in the future.

19. Scroll down the page and click on the National Digital Forecast Database maps. (If your NWS forecast office page does not have the National Digital Forecast Database listing, go back and choose another page that does.) The "Daily View" maps for most weather elements (e.g., temperature) are available at [(*3-*)(*6-*)(*12-*)] hour intervals. Some products are for longer periods and there are tabs for choosing a Weekly View or for Loops.

A *warning* is a statement issued by the National Weather Service indicating that a specified hazardous weather or hydrologic event is imminent or actually occurring. The intention of these warnings is to urge the public to take immediate appropriate action for personal safety. Warnings may be issued for flash floods, hurricanes, severe thunderstorms, tornadoes, or winter storms. Often these warnings are carried by television stations interrupting broadcasts or as text and graphics on-screen or via radio. Special NOAA Weather Radio receivers can be triggered by the local NWS forecast office to signal the issuance of a warning and alert the public.

All weather warnings, watches, advisories, and statements are posted on the maps you have been examining. Browse the national map (*http://www.weather.gov/*) and check on several watches and warnings to familiarize yourself with these products.

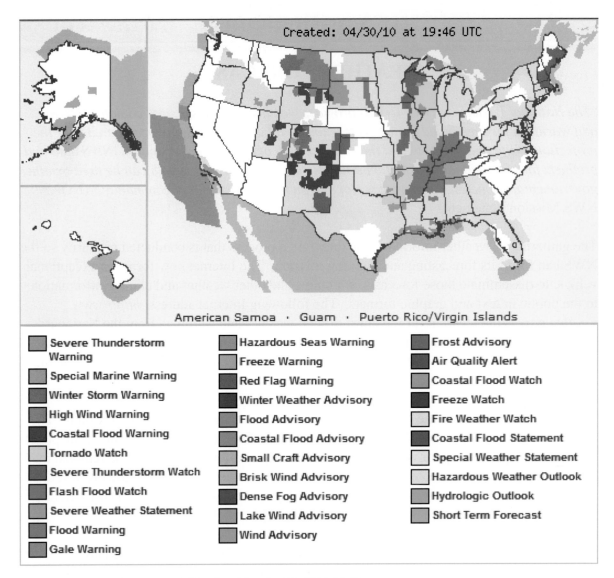

Figure 3. NWS Watches and Warning Map for 30 APR 2010.

Official NWS forecasts are updated several times daily and made available to private meteorological companies and the media. Whether you receive your weather forecast from radio, TV, newspapers or the Internet, the original source for the information is your local NWS forecast office.

Suggestions for further activities: You may wish to call up the website of your local NWS Office. Go to *http://www.wrh.noaa.gov/wrh/forecastoffice_tab.php* to find your nearest NWS Office. Click on the station. From the regional map showing each office's area of responsibility, you can click at your location to obtain the specific forecast for you. Take some time to explore the types of forecast products available.

Investigation
14A:
ATMOSPHERIC OPTICAL PHENOMENA

Objectives:

Solar radiation, consisting of a range of wavelengths, interacts with matter in many ways. The wavelengths from about 0.4 to 0.7 micrometers (μm) are known as visible light because we sense them with our eyes. Visible light also interacts with the air, cloud and precipitation particles, and aerosols in several ways including scattering, reflection, and refraction. These *optical effects* can produce spectacular displays of light and color. These phenomena also give evidence of processes taking place in the atmosphere that may be harbingers of future weather.

After completing this investigation, you should be able to:

- Explain how light interacts with atmospheric water droplets and ice crystals to form rainbows and halos.
- Describe the implications of these optical phenomena for the state of the atmosphere.

Introduction:

As shown in **Figure 1** at the right, when a light ray crosses an interface from one transparent medium to another (such as from air to water), the speed of light changes. If the angle of the light ray is other than perpendicular to the interface, the direction of motion in the new medium is also altered. This change of direction is described by *Snell's law* which states that when a light ray passes into a medium in which the speed of light is slower, the light ray is bent <u>toward</u> the line perpendicular to the interface (Figure 1, ray passing from air to water). If light speeds up when it enters the second medium (Figure 1, ray passing from water to air) a light ray bends <u>away from</u> the perpendicular. Refraction is the bending of a light ray.

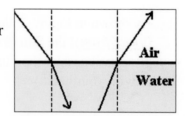

Figure 1.
Bending of light rays
crossing an interface.

For light passing from air (higher speed) into a hexagonal ice crystal (lower speed), the light ray is bent toward the crystal interior. In exiting from the crystal, it is refracted once more, but in reverse fashion. A hexagonal crystal has six rectangular sides, each of which adjoins adjacent sides at a 120-degree angle, and two ends (hexagonal in shape) which are positioned at a 90-degree angle to the sides. A refracted light ray follows one of two paths through a hexagonal ice crystal: through two sides, or a side and an end. The deflection angle of a ray passing through two sides is 22° from its original direction while that through a side and an end is 46° from its original direction.

Figure 2 approximates the orientation of ice crystals that would interact with light rays to form halos of 22° and 46° relative to the observer. The angle is measured where straight lines drawn from the center of the Sun (or Moon) and from the halo meet in the observer's eye. Because ice crystals in the air are more or less randomly oriented, the combination of all

Figure 2.
Light rays bent by ice
crystals (not to scale).

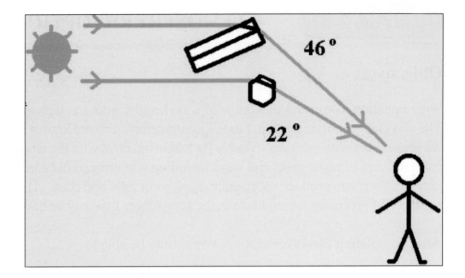

the rays would appear as a circle about the light source. The 22° halo is seen when the light passes in and out of rectangular sides of the ice crystal. The similar process occurs for the 46° halo except the light passes through one side and one end of the crystal. Because there are six sides and only two ends to an ice crystal, probability favors seeing more 22° halos than the 46° kind. The 22° halo is also brighter.

1. As shown in Figure 2, the observer would need to look in the direction [(*toward*) (*away from*)] the Sun to see a halo. (Caution: never look directly at the Sun! Sky observations near the Sun need eye protection or blockage of the Sun's direct rays, such as with your hand.)

2. Halos indicate clouds that are composed of [(*ice crystals*)(*liquid droplets*)].

Figure 3 shows, from the perspective of the observer, the circular 22° and 46° halos about the sun position along with a sun pillar and parhelia (sundogs). A sun pillar is a bright column seen above and below the sun position. The pillar is caused by reflection of the sunlight from the upper and lower surfaces of ice crystals with surfaces oriented horizontally, much like looking at a streetlight through partially opened horizontal venetian blinds. This is best seen near sunrise or sunset when the atmosphere is stable and the larger surfaces of the crystals are horizontally oriented.

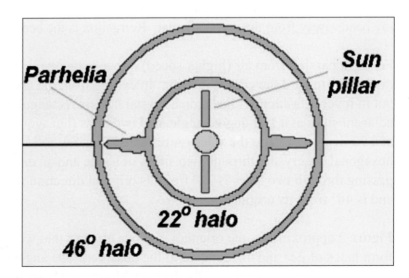

Figure 3.
Halos, parhelia and sun pillar (not to scale).

3. *Parhelia*, commonly called sundogs or mock suns, are bright spots sometimes seen at and just outside the 22° halo circle at the same level as the Sun. The term "sundogs" refer to the dogs that followed the mythological chariot of Mercury, the sun god. Parhelia are primarily formed by the refraction of light passing through hexagonal plate crystals oriented with their relatively large top and bottom six-sided surfaces generally horizontal (falling much like single playing cards which remain horizontal after released from a horizontal position). Parhelia have some coloration caused by refraction of light through the crystals. The distance of parhelia from the Sun increases with increasing solar altitude; at solar altitudes greater than about 60 degrees, parhelia cannot be observed. Stable atmospheric conditions and the presence of horizontal ice crystal surfaces favor the appearance of parhelia and sun pillars, but only **[(*parhelia*)(*sun pillars*)]** require the passing of light through crystals.

4. **Figure 4** suggests the orientation that raindrops would have to the Sun's rays and the observer's location for the formation of a rainbow. The color separation in the primary and secondary rainbows is formed from refraction of the ray both on entering and on leaving a drop. The longer wavelength red light is refracted slightly less than the shorter wavelength violet, resulting in the color separation. The primary rainbow has a single internal reflection of the ray whereas the secondary bow results from **[(*one*)(*two*) (*three*)]** reflections of the ray inside the drop.

5. Rainbows would be seen by looking generally **[(*toward*)(*away from*)]** the Sun.

6. Observing a rainbow can provide weather forecasting hints. A rainbow seen in the morning would be produced by rain falling generally to the **[(*east*)(*west*)]** of the observer.

7. Because weather systems generally move from west to east, this bow-producing rainshower would move **[(*toward*)(*away from*)]** the observer. So, fair weather might be delayed.

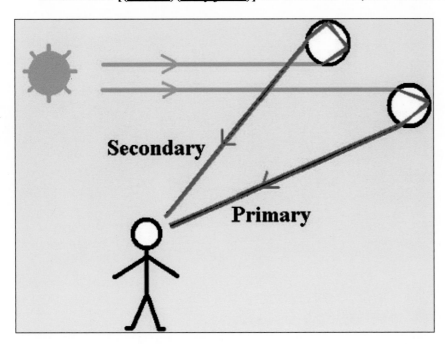

Secondary

Primary

Figure 4.
Rainbow formation (not to scale).

8. In a similar way, a rainbow seen in the afternoon is a harbinger of [(***stormy***)(***clearing***)] weather to follow.

9. Because raindrops have curved surfaces, one [(***would***)(***would not***)] expect to see sun pillars or other surface reflection phenomena from raindrops.

10. Rays from drops produce rainbows in circular arcs about the *antisolar* point, the point opposite the Sun along a line from the Sun through the observer's eyes. The angle measured where straight lines from the antisolar point and from a point on the primary rainbow meet in the observer's eye is about 42°. It is about 50° for the secondary bow. Someone claiming to have seen a rainbow ringing the Sun [(***could be right***) (***would actually have seen a halo***)].

11. Many raindrops must be involved in order for an observer to see a rainbow. A raindrop must be oriented at an angular width of slightly more than 42° to deliver red to the observer's eye while another drop must at slightly less than 41° to deliver violet to the same eye at the same instant. Consequently, red is the [(***outer***)(***inner***)] color in the arc of a primary rainbow. Millions of raindrops fill in the colors of the rainbow to form the bow that can form a circular arc that sometimes stretches from horizon to horizon under proper rain and sunlight conditions.

As directed by your course instructor, complete this investigation by either:

1. *Going to the Current Weather Studies link on the course website, or*
2. *Continuing to the Applications section for this investigation that immediately follows in this Investigations Manual.*

Investigation 14A: Applications

ATMOSPHERIC OPTICAL PHENOMENA

Rainbows are produced from light interacting with water drops in the atmosphere.

12. A view of a primary and a secondary rainbow is shown in **Figure 5**. The anti-solar point, the point opposite the Sun along the line from the Sun through the observer's eyes, is to the lower right corner off the photo. The rainbow forms on a circular arc about that point. With the Sun in the sky and assuming the observer is on a smooth Earth surface with a sea horizon, the anti-solar point is below the horizon so that less than half of the circle on which a rainbow could be formed would be above the horizon. Here the arc of the more colorful primary rainbow separates the part of the image with the lightest background from the part with the darker background. From the inside of the primary rainbow outwards, the colors range from **[(_red to violet_)(_violet to red_)]**.

13. The primary rainbow is formed from the Sun's rays being refracted upon entering a drop, being reflected within the drop, and then refracted again upon leaving the drop. A secondary rainbow would be formed with an additional reflection inside the drop. A faint portion of the secondary rainbow can be seen in the image, well "outside" of the primary bow. The color sequence in the secondary rainbow is reversed from the primary bow, with reddish hues appearing on the **[(_inside_)(_outside_)]** of the secondary rainbow arc.

A faint reddish-purple coloration sometimes seen inside the primary bow is called a supernumerary bow. "Supernumeraries" are formed as diffraction patterns from the primary bow's light rays. The darker area between the primary and secondary bows, known as Alexander's dark band, is produced by the absence of rays being directed in those angles by the reflections.

14. As described earlier in this investigation, light rays may also be reflected and refracted by ice crystals in the atmosphere. Ice crystals interact with light to form halos, as seen in

Figure 5.
A double rainbow view.

Figure 6.
Photograph of a halo

Figure 6. The ring centering on the Sun is a 22-degree halo. The halo is formed by light being refracted through the ice crystals to the observer while the observer looks in the general direction [(*away from*)(*toward*)] the light source.

The atmosphere itself may also refract light to form dramatic optical effects. One may recall the impression that the Sun at sunrise or sunset or the Moon rising or setting seems particularly large. Careful measurement of the angular width of the Sun or Moon shows this is an optical illusion. And, angular measurement of the vertical dimension of the Sun or Moon shows it to be smaller near the horizon than at higher elevations in the sky!

Figure 7, a moonrise, shows the oval shape imparted to the Sun or Moon near the horizon. Rays from the bottom and the top of the Moon's disk are refracted by differing amounts.

15. The image shows the Moon just above the horizon distorted from a circular disc. This departure from a circle results from the Moon's rays traveling through air of varying densities near the Earth's surface. Light from the lower edge of the Moon's disk passes through air of greater density (hence lower speed) than light from the upper edge of the Moon's disk. Consequently, the lower beam undergoes [(*less*)(*more*)] refraction, causing the Moon's lower limb to be elevated more than the upper limb. As the Moon rises higher into the sky and light from across its surface experiences little atmospheric refraction, the Moon's disk appears more and more circular.

Butch Jorgenson
Milwaukee
June 21, 2005

Figure 7.
Flattening of Moon's image.

Also of note in the figure is the reddish color of the Moon. Light reaching the observer (or camera) is coming through a relatively long path of atmosphere where the blues and greens of the moonlight are preferentially absorbed or scattered, leaving red to complete the journey. This attenuation of the shorter wavelengths of visible light is the same process that is responsible for red sunsets and sunrises.

<u>Suggestions for further activities:</u> To view images of rainbows, halos, and other optical phenomena, go to *http://www.atoptics.co.uk/* and *http://www.meteoros.de/indexe.htm*. (Figures 5 and 7 are from NOAA's NWS Forecast Office page at Sullivan, WI.)

ATMOSPHERIC REFRACTION

Objectives:

Light does not always travel in straight lines (even though our minds always assume that the light entering our eyes did just that)! Air may be transparent, but it slows the speed of light passing through it slightly as compared to light's speed in a vacuum. The greater the number of air molecules encountered, the more the light is slowed. When the Sun is close to the horizon, particularly near sunrise or sunset, the Sun's rays entering the atmosphere at a low angle must travel through a relatively long air path. Light rays approaching Earth's surface are progressively slowed as they pass through the higher altitude, less dense air into lower altitude, more dense air. Associated with the slowing is a downward "bending" (refraction) of the light ray. You have observed the effects of this if you have viewed sunsets or sunrises.

After completing this investigation, you should be able to:

• Describe how refraction of light varies with solar altitude.
• Explain how solar refraction affects length of daylight.

Introduction:

You can observe the refraction of light passing from a less dense medium into a more dense medium using an opaque cup, a coin, and water. Place the coin at the bottom of the empty cup and look into the cup at an angle such that you cannot quite see the coin over the edge of the cup (**Figure 1a**). Slowly pour water into the cup, making sure the coin remains in the same position. The coin in the water should appear "magically," as shown in **Figure 1b**. Reflected light from the coin is refracted when it passes from water to air as shown by the solid-line path.

The total path of air through which a light ray from the Sun must pass to an observer varies with the observer's latitude and the time of day. For someone on the equator at local noon on an equinox, the Sun is directly overhead at the zenith and its *solar altitude* is 90 degrees. The

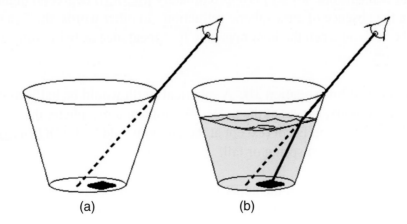

(a)   (b)

Figure 1.
(a) Light ray without refraction and (b) with refraction at a density interface.

Figure 2.
Amount of angular refraction of sunlight as a function of solar altitude.

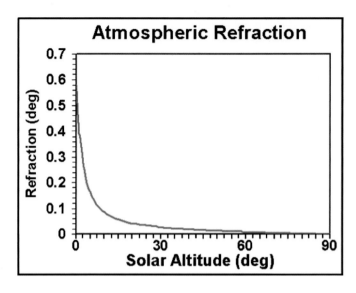

rays of the Sun come straight down and pass through the minimum atmospheric path length. At sunrise or sunset (solar altitude of 0 degrees), the atmospheric path is maximum in length. The amount of refraction or bending of light rays from the Sun due to atmospheric effects can be calculated for the Sun's solar altitude angle. **Figure 2** shows the amount of refraction that occurs (in degrees) for a clear sky at various solar altitudes.

1. Figure 2 shows that the atmospheric refraction of light **[(*increases*)(*decreases*)]** as the solar altitude decreases (approaches the horizon).

2. The maximum amount of atmospheric refraction of a solar ray is **[(*less than one degree*) (*several degrees*)]** when the Sun is on the horizon (solar altitude of 0 degrees).

3. At sunset (or sunrise) with the Sun right on the horizon, the refraction of a Sun's ray is 0.58 degrees. The mean angular diameter of the Sun's visible disk is approximately 0.53 degrees. (Angular diameter is the angle formed at the observer's eye by two lines drawn from opposite edges of the Sun's disk.) Thus, the Sun's position may be distorted by the bending of its rays by slightly more than its own diameter. When very near the horizon, the apparent Sun (the one we see) is approximately **[(*0.5*)(*3*)]** degree(s) higher than it would be in the absence of atmospheric refraction. In other words, the Sun can actually be below the horizon when the bent rays give its appearance as being fully above the horizon!

4. Refer to Figure 3 of Investigation 3B. A spherical Earth would be half sunlit at all latitudes on an equinox, except at the poles. Without an atmosphere, the period of daylight at any latitude on Earth (except at the poles) would be **[(*11*)(*12*)(*13*)]** hours and 0 minutes on that first day of spring or fall.

5. Sunrise or sunset occurs when the upper limb (edge) of the Sun's visible disk is first or last seen under average atmospheric conditions, on a water horizon.

As shown in Figure 6 of Investigation 3B, the daily path of the Sun through the sky varies at different latitudes on Earth. At the equator, the daily path of the Sun is always perpendicular to the horizon. As the latitude increases, the angle between the Sun's path through the daytime sky and the horizon [(**_increases_**)(**_decreases_**)] until it is nearly parallel to the local horizon at the North and South Poles. Thus, the time it takes the Sun to change solar altitude (measured vertically from the horizon) while actually moving steadily along its inclined path will increase with increasing latitude.

In **Table 1**, sunrise and sunset times are given for Kampala, Uganda (0.05 degrees N, 32.4 degrees E), Salem, Oregon (44.9 degrees N, 123.0 degrees W), and Barrow, Alaska (71.3 degrees N, 156.8 degrees W) for 21 March (representative of the first day of spring). For each city calculate the total length of daylight for 21 March and the minutes beyond 12 hours due to atmospheric refraction. [Table values were obtained from the U.S. Naval Observatory website: _http://aa.usno.navy.mil/data/docs/RS_OneYear.php_.]

Table 1: Sunrise/Sunset Times - 21 March			
City	Latitude	Sunrise	Sunset
Kampala	0.1° N	6:54 a.m.	7:01 p.m.
Salem	44.9° N	6:12	6:27
Barrow	71.3° N	7:18	7:53

Note: due to the city's location within its time zone and other astronomical factors, the daylight period is not symmetrical about noon.

6. Kampala's hours and minutes of daylight: 12 hrs [(**_0_**)(**_7_**)(**_15_**)(**_35_**)(**_45_**)] min.
7. Salem's hours and minutes of daylight: 12 hrs [(**_0_**)(**_7_**)(**_15_**)(**_35_**)(**_45_**)] min.
8. Barrow's hours and minutes of daylight: 12 hrs [(**_0_**)(**_7_**)(**_15_**)(**_35_**)(**_45_**)] min.

9. Kampala's minutes of daylight beyond 12 hours: [(**_0_**)(**_7_**)(**_15_**)(**_35_**)(**_45_**)] min.
10. Salem's minutes of daylight beyond 12 hours: [(**_0_**)(**_7_**)(**_15_**)(**_35_**)(**_45_**)] min.
11. Barrow's minutes of daylight beyond 12 hours: [(**_0_**)(**_7_**)(**_15_**)(**_35_**)(**_45_**)] min.

12. The number of minutes of daylight beyond twelve hours on 21 March [(**_increases_**)(**_decreases_**)] with increasing latitude. The additional daylight time is due to the combination of two factors, the orientation of the Sun's path and atmospheric refraction. The smaller the angle of the Sun's path to the horizon, the greater the effect of refraction in lengthening the period of daylight.

As directed by your course instructor, complete this investigation by either:

1. **_Going to the Current Weather Studies link on the course website, or_**
2. **_Continuing to the Applications section for this investigation that immediately follows in this Investigations Manual._**

Investigation 14B: Applications

ATMOSPHERIC REFRACTION

13. The sunrise/sunset data appearing in Table 1 were taken from tables similar to the example shown in Table 2 of this investigation. **Table 2** displays "Rise and Set for the Sun for 2010" for Selawik, Alaska, located at 66.6 degrees N. The sunrise and sunset times are given in local Standard Time on a 24-hour clock; for example, 1842 means 6:42 p.m., local Standard Time. The table indicates that on 20 March 2010 (date of the 2010 spring equinox) the Sun rose at 07:39 and set at **[(*16:19*)(*18:20*)(*19:20*)(*19:57*)]**, local Standard Time.

14. Selawik's period of sunlight on 20 March was 12 hours and **[(*0*)(*3*)(*18*)(*27*)]** minutes.

15. The number of minutes of daylight beyond 12 hours on 20 March at Selawik **[(*fits*)(*does not fit*)]** the pattern of change in the length of daylight with increasing latitude shown in Table 1. [The one day difference in the dates of Table 1 and 2 data can be ignored in the comparison.]

Selawik, at 66.6 degrees N, is located almost exactly on the Arctic Circle (66.5 degrees N). From astronomical considerations based on straight and parallel rays of sunlight striking a spherical Earth, the Arctic Circle (or the Antarctic Circle) should be the lowest latitude at which there would be one continuous 24-hour period of daylight and one continuous 24-hour period of darkness each year.

16. However, according to Table 2, the longest continuous daylight period at Selwik was **[(*1*)(*7*)(*15*)(*33*)(*45*)]** days in length.

17. According to the table, the shortest period of daylight at Selawik occurs on or near the first day of winter in late December and is two hours and **[(*0*)(*7*)(*12*)(*15*)]** minutes long.

18. Atmospheric refraction is the primary cause of the differences between the periods of daylight that might be expected with straight rays of sunlight striking the Earth's surface and the daylight periods calculated by the U.S. Naval Observatory. It was shown earlier in this investigation that when near the horizon the apparent Sun (the one we see) is **[(*higher*)(*lower*)]** than it would be if there were no atmospheric refraction. Thus, over the period of a year the Sun at the Arctic Circle stays continuously above the horizon for considerably more than one day and it never actually goes below the horizon for a continuous 24-hour day.

Suggestions for further activities: The U.S. Naval Observatory provides sunrise and sunset tables for over 22,000 locations in the United States. Go to the following Internet address: *http://www.usno.navy.mil/USNO/astronomical-applications/data-services/rs-one-year-us.*

Scroll down to "Form A – U.S. Cities or Towns". For the current year, specify year and "sunrise/sunset" table and select the state or territory for your hometown. Type in your hometown or where you live in the US, and click on the "Compute Table" button. A table should appear. If the message: *"Unable to find location in our file. Try another location."* is returned, check your spelling or select a nearby larger community.

Print out the table. To print the full 12-month table (you may wish to keep it with your study materials for possible future reference), you must use **landscape** orientation and **8-point** type. Your Internet browser contains options (often called "preferences") that allow you to select the font style and size to use for text files such as this ("fixed font"). If you cannot vary these options yourself, contact your instructor or computer resource person. [Note: Be sure to change your font, size, and orientation back to your original settings after printing out the table.]

Astronomical Applications Dept.
U. S. Naval Observatory
Washington, DC 20392-5420

SELAWIK, ALASKA
Rise and Set for the Sun for 2010

Alaska Standard Time

Location: W160 00, N66 36

Day	Jan. Rise	Jan. Set	Feb. Rise	Feb. Set	Mar. Rise	Mar. Set	Apr. Rise	Apr. Set	May Rise	May Set	June Rise	June Set	July Rise	July Set	Aug. Rise	Aug. Set	Sept. Rise	Sept. Set	Oct. Rise	Oct. Set	Nov. Rise	Nov. Set	Dec. Rise	Dec. Set
	h m	h m	h m	h m	h m	h m	h m	h m	h m	h m	h m	h m	h m	h m	h m	h m	h m	h m	h m	h m	h m	h m	h m	h m
01	1223	1505	1041	1708	0854	1852	0652	2038	0451	2226	0228	0047	****	****	0419	2311	0614	2104	0752	1905	0941	1705	1140	1518
02	1221	1508	1037	1712	0850	1856	0648	2041	0447	2230	0221	0055	****	****	0423	2306	0617	2100	0755	1901	0944	1702	1144	1515
03	1218	1511	1033	1715	0846	1859	0644	2045	0443	2234	0213	0103	****	****	0427	2302	0620	2056	0759	1857	0948	1658	1148	1512
04	1216	1515	1029	1719	0842	1903	0640	2048	0438	2238	0203	0113	****	****	0431	2258	0624	2052	0802	1854	0952	1654	1151	1509
05	1213	1518	1025	1723	0838	1906	0636	2052	0434	2242	****	****	****	****	0435	2254	0627	2048	0805	1850	0956	1650	1155	1506
06	1210	1522	1022	1727	0834	1910	0632	2055	0430	2246	****	****	****	****	0439	2250	0630	2044	0809	1846	1000	1647	1159	1503
07	1207	1526	1018	1731	0830	1913	0628	2058	0426	2250	****	****	****	****	0443	2245	0634	2040	0812	1842	1004	1643	1202	1500
08	1204	1530	1014	1735	0826	1917	0624	2102	0422	2254	****	****	****	****	0447	2241	0637	2036	0815	1838	1008	1639	1206	1458
09	1201	1534	1011	1739	0822	1920	0620	2105	0417	2259	****	****	0209	0121	0451	2237	0640	2032	0819	1834	1011	1635	1209	1455
10	1158	1538	1007	1743	0818	1924	0616	2109	0413	2303	****	****	0220	0111	0455	2233	0644	2028	0822	1830	1015	1632	1212	1453
11	1155	1542	1003	1747	0815	1927	0612	2112	0409	2307	****	****	0228	0103	0458	2229	0647	2024	0826	1826	1019	1628	1216	1451
12	1152	1546	0959	1751	0811	1930	0608	2116	0405	2311	****	****	0235	0056	0502	2225	0650	2020	0829	1822	1023	1624	1219	1449
13	1149	1550	0955	1754	0807	1934	0604	2119	0400	2316	****	****	0242	0050	0506	2221	0653	2016	0832	1818	1027	1620	1221	1447
14	1145	1554	0952	1758	0803	1937	0600	2123	0356	2320	****	****	0248	0044	0510	2217	0657	2012	0836	1815	1031	1617	1224	1446
15	1142	1558	0948	1802	0759	1941	0556	2127	0352	2324	****	****	0254	0038	0514	2213	0700	2008	0839	1811	1035	1613	1226	1444
16	1138	1602	0944	1806	0755	1944	0552	2130	0347	2329	****	****	0300	0033	0517	2208	0703	2004	0843	1807	1039	1609	1228	1443
17	1135	1606	0940	1809	0751	1947	0548	2134	0343	2334	****	****	0305	0028	0521	2204	0706	2000	0846	1803	1043	1606	1230	1442
18	1132	1610	0936	1813	0747	1951	0544	2137	0339	2338	****	****	0310	0023	0525	2200	0710	1956	0850	1759	1047	1602	1232	1442
19	1128	1614	0932	1817	0743	1954	0540	2141	0334	2343	****	****	0315	0018	0528	2156	0713	1952	0853	1755	1052	1559	1233	1441
20	1124	1619	0928	1820	0739	1957	0536	2145	0330	2347	****	****	0320	0013	0532	2152	0716	1949	0857	1751	1056	1555	1234	1442
21	1121	1623	0925	1824	0735	2001	0532	2148	0325	2352	****	****	0325	0008	0535	2148	0719	1945	0900	1748	1100	1552	1235	1442
21														0003										
22	1117	1627	0921	1828	0731	2004	0528	2152	0320	2357	****	****	0330	2359	0539	2144	0723	1941	0904	1744	1104	1548	1235	1443
23	1114	1631	0917	1831	0727	2007	0523	2156	0316	0002	****	****	0335	2350	0543	2140	0726	1937	0907	1740	1108	1545	1235	1443
24	1110	1635	0913	1835	0723	2011	0519	2159	0311	0007	****	****	0339	2345	0546	2136	0729	1933	0911	1736	1112	1541	1235	1445
25	1107	1639	0909	1838	0719	2014	0515	2203	0306	0012	****	****	0344	2341	0550	2132	0732	1929	0915	1732	1116	1538	1234	1446
26	1103	1643	0905	1842	0715	2018	0511	2207	0301	0018	****	****	0349	2336	0553	2128	0736	1925	0918	1728	1120	1534	1234	1448
27	1059	1647	0901	1845	0712	2021	0507	2211	0256	0023	****	****	0353	2332	0556	2124	0739	1921	0922	1724	1124	1531	1232	1450
28	1056	1651	0857	1849	0708	2024	0503	2215	0251	0029	****	****	0357	2328	0600	2120	0742	1917	0926	1721	1128	1528	1231	1453
29	1052	1655			0704	2028	0459	2218	0246	0035	****	****	0402	2323	0603	2116	0746	1913	0929	1717	1132	1524	1230	1455
30	1048	1700			0700	2031	0455	2222	0240	0041	****	****	0406	2319	0607	2112	0749	1909	0933	1713	1136	1521	1228	1458
31	1044	1704			0656	2035			0234	0047			0410	2315	0610	2108			0937	1709			1226	1501

Add one hour for daylight time, if and when in use.

(**** object continuously above horizon)

(---- object continuously below horizon)

Table 2. Daily sunrise and sunset times for Selawik, Alaska (66 degrees, 36 minutes N).

Investigation

15A:

VISUALIZING CLIMATE

Objectives:

Climate is the synthesis of weather conditions, both the average of parameters, generally temperature and precipitation, over a period of time and the extremes in weather. For this reason, much of the information on climate is given in statistical terms. For greater ease of interpretation, these statistical values are often shown in graphs, typically as the magnitude of the average value (or extremes) versus the months of the year. One form of display that shows the relationships between temperature and precipitation during the yearly cycle is the *climograph*.

After completing this investigation, you should be able to:

- Portray the statistical climate values of mean monthly temperature and average monthly precipitation in a graphical form called the climograph.
- Compare temperature and precipitation distributions on climographs from different locations noting similarities and differences.
- Explain how certain climograph patterns can be explained by various climate controls.
- Relate certain patterns of temperature and precipitation to particular climate classification types.

Introduction:

Every place on Earth has climate characteristics that distinguish it from other places. It is desirable to systematically describe these characteristics so that the climates of various locations can be compared. This investigation focuses on climate as described by averages. It is important to remember that by using averages, only a generalized picture of the climate is created. A *climograph* is a commonly used tool to describe the climate of a given place and compare climates in various places. A climograph can be drawn to show monthly mean temperatures and average precipitation totals for a single station through the year on the same graph. Figures 2 - 7 are climographs for six locations in the United States which give examples of major climate types discussed in the Climate Classification, provided at the end of this investigation 15A.

A climograph can provide at a glance the magnitudes and ranges of monthly mean temperatures and average monthly precipitation throughout the year. These statistics are genetically tied to various climate-controlling factors which vary systematically from place to place. By relating distributions of temperature and precipitation to specific controls it is possible to gain a more comprehensive understanding of the causes of the climate in a specific area. Because these controls and the climates which result have significant impact on other elements of the Earth system (vegetation, soils, weathering rates of rocks, etc.), such an understanding has widespread applications. It is also desirable to have a shorthand classification for the major types of recurrent temperature and precipitation patterns so one may be able to generalize about climates up to the global scale.

1. By convention, climographs are usually constructed with time of year displayed horizontally across the base of the graph. The initial letter of the month is listed at mid-month along the bottom, with the precipitation scale along the left side and mean temperature scale on the right side. Mean monthly precipitation totals (rain plus melted snow) are presented as a bar graph. The mean temperature values are plotted as points connected by a curve. In the U.S., climate data are prepared with precipitation in inches and temperatures in degrees Fahrenheit. **Use the data and grid in <u>Figure 1</u> below to make a climograph for Boston, MA. Mark a short horizontal line at mid-month to note the position of the mean total precipitation value of that month and fill in the space below to create a bar. (Use Figures 2 - 7 as a guide.) Place a dot at mid-month at the level denoting the mean monthly temperature value. When all the months are plotted, connect the dots with curved line segments to represent the march of average monthly temperature.**

    Your completed climograph for Boston, MA (Figure 1) shows that the mean monthly temperature rises from near freezing during the winter months (Dec, Jan, and Feb) to means around 70 °F during the summer months (June, Jul, and Aug) and then falls as winter approaches. The observed temperatures from which the means are computed result mainly from the seasonal swing of solar heating, which in turn is largely determined by latitude. As a general rule, the higher the latitude the lower the winter season temperatures. The lowest mean monthly temperature in Boston occurs in **[(*<u>January</u>*)(*<u>December</u>*)]**.

2. This minimum monthly temperature **[(*<u>is</u>*)(*<u>is not</u>*)]** within a month or so of the time of minimum solar heating in a mid-latitude, Northern Hemisphere location.

3. Where solar heating varies significantly from the winter to summer solstices, the range of temperatures, indicated by the amplitude of the temperature curve on a climograph, is relatively great. Where the amplitude is relatively small, the seasonal temperature contrast is also small. Examine the temperature curve on the climograph for Hilo, HI **(Figure 2)**. The range of mean monthly temperatures for Hilo is about **[(*<u>20</u>*)(*<u>5</u>*)(*<u>30</u>*)]** Fahrenheit degrees.

4. From the shape of the curve and range of temperature, it is evident that Hilo experiences relatively **[(*<u>little</u>*)(*<u>significant</u>*)]** variation in solar heating through the course of a year.

5. The highest mean monthly temperature in Hilo occurs in August. This temperature is about **[(*<u>76</u>*)(*<u>86</u>*)]** °F.

6. The lowest mean monthly temperature is about **[(*<u>71</u>*)(*<u>81</u>*)]** °F. in both January and February.

7. These temperatures suggest that Hilo is a **[(*<u>high</u>*)(*<u>low</u>*)]** latitude location.

Month	Temp.(F)	Precip.(in)
J	29.3	3.92
F	31.5	3.30
M	38.9	3.85
A	48.3	3.60
M	58.5	3.24
J	68.0	3.22
J	73.9	3.06
A	72.3	3.37
S	64.7	3.47
O	54.1	3.79
N	44.9	3.98
D	34.8	3.73

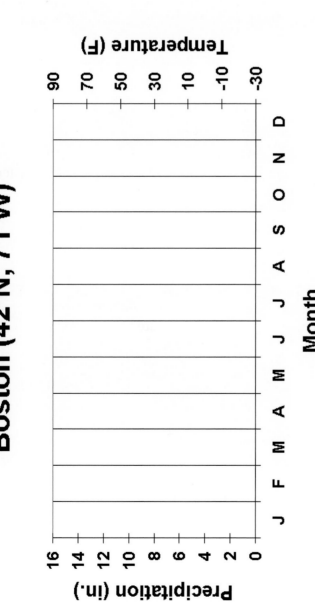

Boston (42 N, 71 W)

Figure 1.
Humid Continental Climate (Dfa) - Boston.

8. The month of occurrence of the highest mean temperature suggests that Hilo is located in the [(***Northern***)(***Southern***)] Hemisphere.

9. Temperature and temperature range can also be influenced by large bodies of water (ocean or large lake). Generally speaking, a maritime influence will moderate temperatures in places that would normally be colder in winter and warmer in summer based on latitude alone. The seasonal range in temperatures is likely to be less due to a maritime influence; that is, the temperature curve on the climograph will exhibit less amplitude. In continental locations or locations downwind of large land masses temperatures tend to be higher in summer and lower in winter. As a result, the seasonal temperature range will be far greater than for places surrounded by or downwind of a large water body. It is likely that Hilo's relatively low annual temperature range [(***is***) (***is not***)] also moderated by the surrounding Pacific Ocean.

10. Examine the climograph for Fairbanks, AK (**Figure 6**). The highest mean monthly temperature is [(***62***)(***82***)] °F.

11. The lowest monthly mean temperature for Fairbanks is [(***−10***)(***10***)] °F.

12. These temperatures suggest that Fairbanks is a [(***high***)(***low***)] latitude location.

13. The average monthly temperatures in Fairbanks cover a range of about [(***70***)(***50***)(***30***)] Fahrenheit degrees.

14. This range of temperatures suggests that Fairbanks has a [(***continental***)(***maritime***)] climate.

15. Compare the annual temperature range for Boston, on the Atlantic coast, (**Figure 1**) and Seattle, WA, near the Pacific coast, (**Figure 5**). Seattle's annual temperature range is [(***greater than***)(***less than***)] that of Boston.

16. These cities are at approximately the same latitude and both are located near the coast. The climate control causing the difference in annual temperature range is the influence of the prevailing westerly wind at both locations. For Seattle, the temperature range is influenced mainly by the [(***ocean***)(***continent***)] which is upwind and in Boston by winds blowing from the continent.

17. Climate classification systems allow climate differences and similarities to be expressed in a "shorthand" form. The broad-scale climate boundaries in the *Köppen* climate classification system (see Climate Classification at the end of this investigation) are based on patterns in annual and monthly mean temperature and precipitation, which closely correspond to the limits of vegetative communities. The major classifications of *Tropical Humid* (A), *Subtropical* (C), *Snow Forest* (D), and *Polar* (E) are based on temperature; the group *Dry* (B) is based on precipitation; and the group *Highland* (H) applies to mountainous regions. Temperatures for both Fairbanks and Boston place them in the [(***Tropical Humid (A)***)(***Subtropical (C)***)(***Snow Forest (D)***)(***Polar (E)***)] classification.

18. The second letter of the Boston (Figure 1) and Fairbanks (Figure 6) classification corresponds to seasonal precipitation regimes with an "f" signifying year-round precipitation. According to their climographs and climate classifications, Boston and Fairbanks have [(*similar*)(*very different*)] seasonal precipitation regimes.

19. Arid and Semiarid climates can be caused by several climate controls. Locations on the eastern side of planetary-scale, persistent high pressure systems, such as those occurring around 30 degrees N in the Atlantic and Pacific, experience subsiding air, which inhibits cloud formation and precipitation. The west side of such systems, by contrast, tend to be humid. The cause of dryness in Tucson AZ, for example, is due mainly to its position [(*east*)(*west*)] of a subtropical high pressure system which persists off the southwest U.S. coast in the Pacific.

20. Atlanta, GA at about the same latitude as Tucson, but in the southeastern United States, is humid because it is located [(*east*)(*west*)] of such a high pressure system in the Atlantic.

21. Dry or wet conditions can also be caused by location upwind or downwind of a mountain range. Areas to the lee of high mountains tend to be dry because of the "wringing out" of moisture on the wet, windward slopes (due to orographic lifting, cooling and condensation) and the compressional warming of air which occurs as the air descends on the leeward slopes. The atmospheric stability caused by cold ocean currents offshore can also prevent precipitation by stabilizing the air and inhibiting convection. Instability can occur if ocean currents are warm. The dryness of Tucson, which is downwind of the Coastal Ranges and the cold California Current, is [(*probably*)(*not likely*)] drier because of the influence of mountains and ocean currents.

As directed by your course instructor, complete this investigation by either:

1. *Going to the Current Weather Studies link on the course website, or*
2. *Continuing to the Applications section for this investigation that immediately follows in this Investigations Manual.*

Hilo (19 N, 155 W)

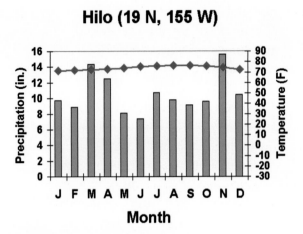

Figure 2.
Tropical Wet Climate (Af) - Hilo.

Tucson (32 N, 111 W)

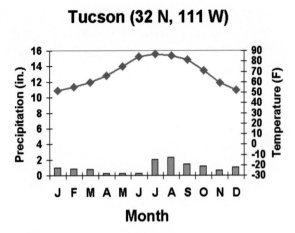

Figure 3.
Subtropical Desert Climate (BWh) - Tucson.

Atlanta (33 N, 84 W)

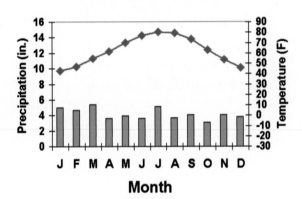

Figure 4.
Subtropical Humid Climate (Cfa) - Atlanta.

Seattle (47 N, 122 W)

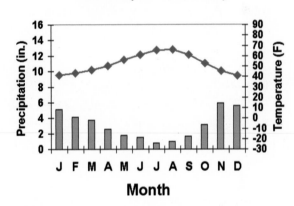

Figure 5.
Marine West Coast Climate (Cfb) - Seattle.

Fairbanks (64 N, 147 W)

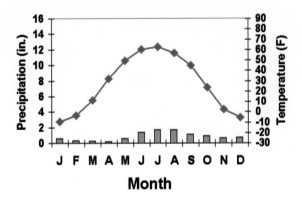

Figure 6.
Subarctic Climate (Dfc) - Fairbanks.

Barrow (71 N, 156 W)

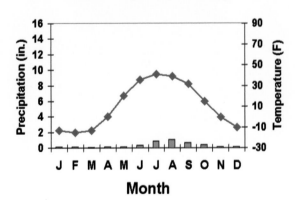

Figure 7.
Polar Tundra Climate (Et) - Barrow.

Investigation 15A: Applications

VISUALIZING CLIMATE

22. The **Figure 8** world map showing the Koeppen (or Köppen) climate classification demonstrates the actual application of climatic controls. For example, on this Mercator projection map, horizontal lines (if they were drawn) would represent constant latitudes. Therefore, at similar latitudes across the Eurasian land mass, Europe to the west is shown in purple, indicating a temperate climate while eastern Asia is in yellow indicating a cold type climate. The climate control primarily at work in these local climate types is the prevailing wind circulation in relation to **[(*elevation*)(*proximity to large bodies of water*) (*Earth's surface characteristics*)]**.

23. The region of Tibet in south central Asia is shown in the greenish-brown of a polar type climate although it is surrounded by dry or temperate climates. This classification is most likely the result of Tibet's **[(*elevation*)(*proximity to large bodies of water*)(*Earth's surface characteristics*)]**

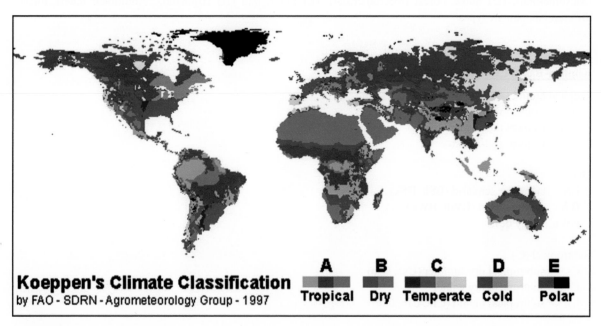

Figure 8.
Koeppen's Climate classification. *(Adapted from UN Food and Agriculture Organization, Sustainable Development website.)*

Suggestions for further activities: You can make your own climographs with monthly average temperatures and precipitation totals from *http://www.worldclimate.com*. Monthly and annual values are provided in both English and metric units. By inputting data to spreadsheet software, graphing can be easily accomplished allowing comparisons among stations.

CLIMATE CLASSIFICATION

- Tropical Humid Climates
- Dry Climates
- Subtropical Climates
- Snow Forest Climates
- Polar Climates
- Highland Climates

One of the most widely used climate classification systems was designed by German climatologist and plant geographer Wladimir Köppen (1846-1940) and subsequently modified by his students R. Geiger and W. Pohl. The Köppen system is an empirical approach to organizing Earth's myriad of climate types. Recognizing that indigenous vegetation is a natural indicator of regional climate, Köppen and his students looked for patterns in annual and monthly mean temperature and precipitation, which closely correspond to the limits of vegetative communities thereby revealing broad-scale climatic boundaries throughout the world. Records of annual and monthly mean temperature and precipitation are sufficiently long and reliable in many parts of the world that they serve as a good first approximation of climate. Since its introduction in the early 1900s, Köppen's climate classification has undergone numerous and substantial revisions by Köppen himself and by other climatologists and has had a variety of applications.

As shown in the Table below, the Köppen climate classification system identifies six main climate groups; four are based on temperature, one is based on precipitation, and one applies to mountainous regions. Köppen's scheme uses letters to symbolize major climatic groups: (A) Tropical Humid, (B) Dry, (C) Subtropical (Mesothermal), (D) Snow Forest (Microthermal), (E) Polar, and (H) Highland. Additional letters further differentiate climate types.

TABLE
Köppen-Based Climate Classification

Tropical humid (A)
   Af   tropical wet
   Am tropical monsoon
   Aw tropical wet-and-dry

Dry (B)
   BS   steppe or semiarid (BSh, BSk)
   BW arid or desert (BWh, BWk)
   BWn foggy desert

Subtropical (C)
   Cs   subtropical dry summer (Csa, Csb)
   Cw subtropical dry winter
   Cf   subtropical humid (Cfa, Cfb, Cfc)

Snow forest (D)
   Dw dry winter (Dwa, Dwb, Dwc, Dwd)
   Df   year-round precip. (Dfa, Dfb, Dfc, Dfd)
   Ds   dry summer

Polar (E)
   Et   tundra
   Ef   ice cap

Highland (H)

Tropical Humid Climates

Tropical humid climates (A) constitute a discontinuous belt straddling the equator and extending poleward to near the Tropic of Cancer in the Northern Hemisphere and the Tropic of Capricorn in the Southern Hemisphere. Mean monthly temperatures are high and exhibit little variability throughout the year. The mean temperature of the coolest month is no lower than 18 °C (64 °F), and there is no frost. The temperature contrast between the warmest and coolest month is typically less than 10 Celsius degrees (18 Fahrenheit degrees). In fact, the diurnal (day-to-night) temperature range generally exceeds the annual temperature range. This monotonous air temperature regime is the consequence of consistently intense incoming solar radiation associated with a high maximum solar altitude and little variation in the period of daylight throughout the year.

Although tropical humid climate types are not readily distinguishable on the basis of temperature, important differences occur in precipitation regime. Tropical humid climates are subdivided into tropical wet (Af), tropical monsoon (Am), and tropical wet-and-dry (Aw). Although these climate types generally feature abundant annual rainfall, more than 100 cm (40 in.) on average, their rainy seasons differ in length and, in the case of Am and Aw, there is a pronounced dry season and wet season. In tropical wet climates, the yearly average rainfall of 175 to 250 cm (70 to 100 in.) supports the world's most luxurious vegetation. Tropical rainforests occupy the Amazon Basin of Brazil, the Congo Basin of Africa, and the islands of Micronesia. For the most part, rainfall is distributed uniformly throughout the year, although some areas experience a brief (one or two month) dry season. Rainfall occurs as heavy downpours in frequent thunderstorms triggered by local convection and the intertropical convergence zone (ITCZ). Convection is largely controlled by solar radiation and rainfall typically peaks in midafternoon, the warmest time of day. Because the water vapor concentration is very high, even the slightest cooling at night leads to saturated air and the formation of dew or radiation fog, giving these regions a sultry, steamy appearance.

Tropical monsoon (Am) climates feature a seasonal rainfall regime with extremely heavy rainfall during several months and a lengthy dry season. The principal control for these climates involves seasonal shifts in wind from land to sea, typified by the Asian monsoon. During the low-sun season, high air pressure over the Asian continent causes dry air to flow southward into parts of Southeast Asia and India. During the high-sun season, low air pressure covers the Tibetan Plateau and the winds reverse direction, advecting moisture inland from over the Indian Ocean. Local convection, orographic lifting, and shifts of the ITCZ combine to deluge the land with torrential rains. Am climates also occur in western Africa and northeastern Brazil.

For the most part, tropical wet-and-dry climates (Aw) border tropical wet climates (Af) and are transitional to subtropical dry climates in a poleward direction. Aw climates support the savanna, tropical grasslands with scattered deciduous trees. Summers are wet and winters are dry, with the dry season lengthening poleward. This marked seasonality of rainfall is linked to shifts of the intertropical convergence zone (ITCZ) and semipermanent subtropical anticyclones, which follow the seasonal excursions of the sun. In summer (*high-sun* season), surges of the ITCZ trigger convective rainfall; in winter (*low-sun* season), the dry eastern flank of the subtropical anticyclones dominates the weather.

The annual mean temperature in Aw climates is only slightly lower, and the seasonal temperature range is only slightly greater, than in the tropical wet climates (Af). The diurnal temperature range varies seasonally, however. In summer, frequent cloudy skies and high humidity suppress the diurnal temperature range by reducing both solar heating during the day and radiational cooling at night. In winter, on the other hand, persistent fair skies have the opposite effect on radiational heating and cooling and increase the diurnal temperature range. Cloudy, rainy summers plus dry winters also mean that the year's highest temperatures typically occur toward the close of the dry season in late spring.

Dry Climates

Dry climates (B) characterize those regions where average annual potential evaporation exceeds average annual precipitation. *Potential evaporation* is the quantity of water that would vaporize into the atmosphere from a surface of fresh water during long-term average weather conditions. Air temperature largely governs the rate of evaporation so it is not possible to specify some maximum rainfall amount as the criterion for dry climates.

Rainfall is not only limited in B climates but also highly variable and unreliable. As a general rule, the lower the mean annual rainfall, the greater is its variability from one year to the next.

Earth's dry climates encompass a larger land area than any other single climate grouping. Perhaps 30% of the planet's land surface, stretching from the tropics into midlatitudes, experiences a moisture deficit of varying degree. These are the climates of the world's deserts and steppes, where vegetation is sparse and equipped with special adaptations that permit survival under conditions of severe moisture stress. Based on the degree of dryness, we distinguish between two dry climate types: steppe or semiarid (BS) and arid or desert (BW). Steppe or semiarid climates are transitional between more humid climates and arid or desert climates. Mean annual temperature is latitude dependent, as is the range in variation of mean monthly temperatures through the year. Hence, a distinction is made between warm dry climates of tropical latitudes (BSh and BWh) and cold, dry climates of higher latitudes (BSk and BWk).

Dryness is the consequence of subtropical anticyclones, cold surface ocean currents, or the rain shadow effect of high mountain ranges. Subsiding stable air on the eastern flanks of subtropical anticyclones gives rise to tropical dry climates (BSh and BWh). These huge semipermanent pressure systems, centered over the ocean basins, dominate the weather year-round near the Tropics of Cancer and Capricorn. Consequently, dry climates characterize North Africa eastward to northwest India, the southwestern United States and northern Mexico, coastal Chile and Peru, southwest Africa, and much of the interior of Australia.

Although persistent and abundant sunshine is generally the rule in dry tropical climates, there are some important exceptions. Where cold ocean waters border a coastal desert, a shallow layer of stable marine air drifts inland. The desert air thus features high relative humidity, persistent low stratus clouds and fog, and considerable dew formation. Examples are the Atacama Desert of Peru and Chile, the Namib Desert of southwest Africa, and portions of the coastal Sonoran Desert of Baja California and stretches of the coastal Sahara Desert of northwest Africa. These anomalous foggy desert climates are designated BWn.

Cold, dry climates of higher latitudes (BWk and BSk) are situated in the rain shadows of great mountain ranges. They occur primarily in the Northern Hemisphere, to the lee of the Sierra Nevada and Cascade ranges in North America and the Himalayan chain in Asia. Because these dry climates are at higher latitudes than their tropical counterparts, mean annual temperatures are lower and the seasonal temperature contrast is greater. Anticyclones dominate winter, bringing cold and dry conditions, whereas summers are hot and generally dry. Scattered convective showers, mostly in summer, produce relatively meager precipitation.

Subtropical Climates

Subtropical climates are located just poleward of the Tropics of Cancer and Capricorn and are dominated by seasonal shifts of subtropical anticyclones. There are three basic climate types: subtropical dry summer (or *Mediterranean*) (Cs), subtropical dry winter (Cw), and subtropical humid (Cf), which receive precipitation throughout the year.

Mediterranean climates occur on the western side of continents between about 30 and 45 degrees latitude. In North America, mountain ranges confine this climate to a narrow coastal strip of California. Elsewhere, Cs climates rim the Mediterranean Sea and occur in portions of extreme southern Australia. Summers are dry because at that time of year Cs regions are under the influence of stable subsiding air on the eastern flanks of the semi-permanent subtropical highs. Equatorward shift of subtropical highs in autumn allows extra-tropical cyclones to migrate inland, bringing moderate winter rainfall. Mean annual precipitation varies greatly-ranging from 30 to 300 cm (12 to 80 in.) with the wettest winter month typically receiving at least three times the precipitation of the driest summer month.

Although Mediterranean climates exhibit a pronounced seasonality in precipitation (dry summers and wet winters), the temperature regime is quite variable. In coastal areas, cool onshore breezes prevail, lowering the mean annual temperature and reducing seasonal temperature contrasts. Well inland, however, away from the ocean's moderating influence, summers are considerably warmer; hence, inland mean annual temperatures are higher and seasonal temperature contrasts are greater than in coastal Cs localities. Climatic records of coastal San Francisco and inland Sacramento, CA illustrate the contrast in temperature regime within Cs regions. Although the two cities are separated by only about 145 km (90 mi.), the climate of Sacramento is much more continental (much warmer summers and somewhat cooler winters) than that of San Francisco. The warm climate subtype is designated Csa and the cooler subtype is Csb.

Subtropical dry winter climates (Cw) are transitional between Aw and BS climates and located in South America and Africa between 20 and 30 degrees S. Cw climates also occur between the Aw and H climates of the Himalayas and Tibetan plateau and between the BS and Cfa climates of Southeast and East Asia. Northward shift of the subtropical high pressure systems is responsible for the dry winter in South America and Africa. The narrowness of the two continents between 20 and 30 degrees S means a relatively strong maritime influence and dictates against extreme dryness. In spring, subtropical highs shift southward and rains return. In Asia, winter dryness is caused by winds radiating outward from the massive cold Siberian high. As the continent warms in spring, the Siberian high weakens and eventually is replaced by low pressure. Moist winds then flow inland bringing summer rains. Mean annual precipitation in Cw climates is in the range of 75 to 150 cm (30 to 60 in.).

Subtropical humid climates (Cf) occur on the eastern side of continents between about 25 and 40 degrees latitude (and even more poleward where the maritime influence is strong). Cfa climates are the most important of the Cf climate subtypes in terms of land area and number of people impacted. Cfa climates are situated primarily in the southeastern United States, a portion of southeastern South America, eastern China, southern Japan, on the extreme southeastern coast of South Africa, and along much of the east coast of Australia. These climates feature abundant precipitation (75 to 200 cm, or 30 to 80 in., on average annually), which is distributed throughout the year. In summer, Cfa regions are dominated by a flow of sultry maritime tropical air on the western flanks of the subtropical anticyclones. Consequently, summers are hot and humid with frequent thunderstorms, which can produce brief periods of substantial rainfall. Hurricanes and tropical storms contribute significant rainfall (up to 15% to 20% of the annual total) to some North American and Asian Cfa regions, especially from summer through autumn. In winter, after the subtropical highs shift toward the equator, Cfa regions come under the influence of migrating extratropical cyclones and anticyclones.

In Cfa localities, summers are hot and winters are mild. Mean temperatures of the warmest month are typically in the range of 24 to 27 °C (75 to 81 °F). Average temperatures for the coolest months typically range from 4 to 13 °C (39 to 55 °F). Subfreezing temperatures and snowfalls are infrequent.

A strong maritime influence is responsible for the cool summers and mild winters of Cfb climates. These climates occur over much of Northwest Europe, New Zealand, and portions of southeastern South America, southern Africa, and Australia.   The coldest subtype, the Cfc, is relegated to coastal areas of southern Alaska, Norway, and the southern half of Iceland. Cfb and Cfc climates are relatively humid with mean annual precipitation ranging between 100 and 200 cm (40 and 80 in.).

Snow Forest Climates

Snow forest climates (D) occur in the interior and to the leeward sides of large continents. The name emphasizes the link between biogeography and the Köppen climate classification system. These climates feature cold snowy winters (except for the Dw subtype in which the winter is dry) and occur only in the Northern Hemisphere. Snow forest climates are subdivided according to seasonal precipitation regimes with Df climates experiencing year-round precipitation whereas Dw climates have a dry winter. D climates with dry summers (Ds) are rare and small in extent. Additional distinction is made between warmer subtypes (Dwa, Dfa, Dwb, and Dfb) and colder subtypes (Dwc, Dfc, Dwd, Dfd).

The warmer subtypes, sometimes termed *temperate continental*, have warm summers (mean temperature of the warmest month greater than 22 °C or 71 °F) and cold winters. They are located in Eurasia, the northeastern third of the United States, southern Canada, and extreme eastern Asia. Continentality increases inland with maximum temperature contrasts between the coldest and warmest months as great as 25 to 35 Celsius degrees (45 to 63 Fahrenheit degrees). The southerly Dfa climates have cool winters and warm summers and the more northerly Dfb climates have cold winters and mild summers. The freeze-free period varies in length from 7 months in the south to only 3 months in the north. The weather in these regions is very changeable and dynamic because these areas are swept by extra-tropical cyclones and anticyclones and by surges of contrasting air masses. Polar front cyclones dominate winter, bringing episodes of light to moderate frontal precipitation. These storms are followed by incursions of dry polar and arctic air masses. In summer, cyclones are weak and infrequent as the principal storm track shifts poleward. Summer rainfall is mostly convective, and locally amounts can be very heavy in severe thunderstorms and mesoscale convective complexes (MCCs). Although precipitation is distributed rather uniformly throughout the year, most places experience a summer maximum.

In northern portions, winter snowfall becomes an important factor in the climate. Mean annual snowfall and the persistence of a snow cover increase northward. Because of its high albedo for solar radiation and its efficient emission of infrared, a snow cover chills and stabilizes the overlying air. For these reasons, a snow cover tends to be self-sustaining; once established in early winter, an extensive snow cover tends to persist.

Moving poleward, summers get colder and winters are bitterly cold. These so-called *boreal climates* (Dfc, Dfd, Dwc, Dwd) occur only in the Northern Hemisphere as an east-west band between 50 to 55 degrees N and 65 degrees N. It is a region of extreme continentality and very low mean annual temperature. Summers are short and cool, and winters are long and bitterly cold. Because midsummer freezes are possible, the growing season is precariously short. Both continental polar (cP) and arctic (A) air masses originate here, and this area is the site of an extensive coniferous (boreal) forest. In summer, the mean position of the leading edge of arctic air (the arctic front) is located along the northern border of the boreal forest. In winter, the mean position of the arctic front is situated along the southern border of the boreal forest.

Weak cyclonic activity occurs throughout the year and yields meager annual precipitation (typically less than 50 cm, or 20 in.). Convective activity is rare. A summer precipitation maximum is due to the winter dominance of cold, dry air masses. Snow cover persists throughout the winter and the range in mean temperature between winter and summer is among the greatest in the world.

Polar Climates

Polar climates (E) occur poleward of the Arctic and Antarctic circles. These boundaries correspond roughly to localities where the mean temperature for the warmest month is 10 °C (50 °F). These limits also approximate the tree line, the poleward limit of tree growth. Poleward are tundra and the Greenland and Antarctic ice sheets. A distinction is made between tundra (Et) and ice cap (Ef) climates, with the dividing criterion being 0 °C (32 °F) for the mean temperature of the warmest month. Vegetation is sparse in Et regions and almost nonexistent in Ef areas.

Polar climates are characterized by extreme cold and slight precipitation, which falls mostly in the form of snow (less than 25 cm, or 10 in., melted, per year). Greenland and Antarctica could be considered deserts for lack of significant precipitation, despite the presence of large ice sheets. Although summers are cold, the winters are so extremely cold that polar climates feature a marked seasonal temperature contrast. Mean annual temperatures are the lowest of any place in the world.

Highland Climates

Highland Climates (H) encompass a wide variety of climate types that characterize mountainous terrain. Altitude, latitude, and exposure are among the factors that shape a complexity of climate types. For example, temperature decreases rapidly with increasing altitude and windward slopes tend to be wetter than leeward slopes. Climate-ecological zones are telescoped in mountainous terrain. That is, in ascending several thousand meters of altitude, we encounter the same bioclimatic zones that we would experience in traveling several thousand kilometers of latitude. As a general rule, every 300 m (980 ft) of elevation corresponds roughly to a northward advance of 500 km (310 mi).

Investigation
15B:

LOCAL CLIMATE DATA

Objectives:

Climate data are extremely useful for numerous purposes. Farmers use their knowledge of weather and climate over a long period to determine what crops to plant and for guidance on when to plant and when to harvest. Utilities use climate data for planning production and distribution of energy supplies and the reallocation among types of such supplies. The building industry uses climate data in the design of structures, including their necessary strength, heating and cooling energy requirements, and the associated building codes that regulate them. People look to climate data as they plan future events or activities (e.g., outdoor gatherings, vacations, and sporting events). These are just a few of the multitude of uses for climate data.

In the U.S., weather data are gathered by NOAA's National Weather Service offices and other organizations and compiled at state, regional, and national centers for distribution to users. NOAA's National Climatic Data Center in Asheville, NC is responsible for compiling U.S. data as well as being a depository for much worldwide data on weather and the environment. This information, in turn, is made available to users in print, CD-ROM, and online electronic formats.

After completing this investigation, you should be able to:

- Interpret information appearing in *Local Climatic Data, Annual Summary with Comparative Data* based on weather data collected at a local National Weather Service office.
- Determine how to access archived climate data from National Climatic Data Center (NCDC).

Introduction:

1. A basic publication of the National Climatic Data Center based on weather data from local National Weather Service (NWS) offices is the *Local Climatological Data* (LCD). It is published in monthly and annual summaries. LCDs are published for about 275 NWS observing sites. Portions of the *LCD, Annual Summary with Comparative Data*, for Grand Island, Nebraska (GRI) for the year 2009 are used in this investigation. Grand Island is located near the geographic center of the coterminous United States.

    Examine the temperature graph appearing on the report's front page **(Figure 1)**. Daily temperature ranges are plotted as vertical lines on the graph. The top end of each line signifies the maximum daily temperature and the bottom end reports the day's minimum temperature. Assuming that frost occurs if the temperature falls to 32 °F or lower, the approximate date of the last spring frost in 2009 was about **[(*6*)(*19*)(*28*)]** April.

2. The date of the year's first fall frost was about **[(*8*)(*17*)(*29*)]** September.

2009
LOCAL CLIMATOLOGICAL DATA
ANNUAL SUMMARY WITH COMPARATIVE DATA

GRAND ISLAND,
NEBRASKA (KGRI)

ISSN 0198-3105

Daily Max/Min Temperature

Legend: — Normal Max — Normal Min — Freezing ◆ Max/Min

Daily Precipitation

Daily Station Pressure

I CERTIFY THAT THIS IS AN OFFICIAL PUBLICATION OF THE NATIONAL OCEANIC AND ATMOSPHERIC ADMINISTRATION, AND IS COMPILED FROM RECORDS ON FILE AT THE NATIONAL CLIMATIC DATA CENTER.

NATIONAL	NATIONAL	NATIONAL	
OCEANIC AND	ENVIRONMENTAL SATELLITE, DATA	CLIMATIC DATA CENTER	DIRECTOR
ATMOSPHERIC ADMINISTRATION	AND INFORMATION SERVICE	ASHEVILLE, NORTH CAROLINA	NATIONAL CLIMATIC DATA CENTER

Figure 1. Local Climatological Data, Annual Summary, Grand Island, NE 2009, cover.

Examine page 2 of the *LCD, Annual Summary* (**Figure 2**) entitled, **METEOROLOGICAL DATA FOR 2009"** and answer the following questions about Grand Island, NE:

3. What month had the lowest average temperature ("Average Dry Bulb")? **[(*December*) (*January*)(*February*)]**

4. What was the average temperature that month? **[(*9.9*)(*18.3*)(*21.6*)(*23.5*)]** °F.

5. What month had the highest average temperature? **[(*June*)(*July*)(*August*)]**

6. What was that temperature? **[(*72.5*)(*79.8*)(*84.4*)(*98.4*)]** °F

7. How many days in the year had temperatures of 90 °F or higher? **[(*9*)(*14*)(*20*)(*45*)]**

8. How many had temperatures of 0 °F or lower? **[(*8*)(*13*)(*29*)(*45*)]**

9. On how many days in the year were thunderstorms reported? **[(*6*)(*17*)(*29*)(*37*)]**

10. The strongest gust of wind, as reported in the Maximum 3-Second Wind category was a speed of **[(*41*)(*49*)(*54*)(*66*)]** mph.

11. This occurred during 2009 in the month of **[(*March*)(*June*)(*September*)(*November*)]**.

12. The total precipitation for the year (Water Equivalent: Total) was about **[(*9*)(*14*)(*26*)(*43*)]** inches.

13. The total number of days when there was at least a trace of precipitation (equal to or greater than 0.01 in.) was **[(*9*)(*29*)(*79*)(*109*)]**.

14. The greatest 24-hour snowfall during the year was **[(*6.6*)(*9.7*)(*14.7*)(*23.6*)]** inches.

15. This occurred during the month of **[(*December*)(*January*)(*February*)(*March*)]**.

METEOROLOGICAL DATA FOR 2009
GRAND ISLAND (KGRI)

LATITUDE: 40 ° 57'N LONGITUDE: -98 ° 18'W ELEVATION (FT): GRND: 1850 BARO: 1844 TIME ZONE: CENTRAL (UTC -6) WBAN: 14935

ELEMENT	JAN	FEB	MAR	APR	MAY	JUN	JUL	AUG	SEP	OCT	NOV	DEC	YEAR
TEMPERATURE °F													
MEAN DAILY MAXIMUM	39.3	43.6	51.8	61.9	74.8	79.3	84.4	82.9	75.0	54.2	56.8	26.7	60.9
HIGHEST DAILY MAXIMUM	59	65	80	89	91	98	98	95	84	71	76	58	98
DATE OF OCCURRENCE	31+	24	16	23	31	23	24	07	08+	19+	06	01	JUL 24
MEAN DAILY MINIMUM	13.8	18.4	25.1	36.4	49.9	59.6	60.6	59.9	51.9	35.0	30.6	9.9	37.6
LOWEST DAILY MINIMUM	-6	-3	-1	14	38	49	50	45	32	24	17	-8	-8
DATE OF OCCURRENCE	24	14	01	07	16+	03	18	30	29	10	18	15	DEC 15
AVERAGE DRY BULB	26.6	31.0	38.5	49.2	62.4	69.5	72.5	71.4	63.5	44.6	43.7	18.3	49.3
MEAN WET BULB	22.4	26.6	33.0	42.2	54.0	63.1	64.8	64.1	56.7	40.4		16.9	
MEAN DEW POINT	16.0	19.7	25.5	34.3	46.8	59.1	60.3	60.1	52.4	35.9	31.9	12.5	37.9
NUMBER OF DAYS WITH:													
MAXIMUM >= 90°	0	0	0	0	3	6	4	7	0	0	0	0	20
MAXIMUM <= 32°	6	5	4	0	0	0	0	0	0	0	0	25	40
MINIMUM <= 32°	31	25	23	12	0	0	0	0	1	12	20	31	155
MINIMUM <= 0°	3	2	1	0	0	0	0	0	0	0	0	7	13
H/C													
HEATING DEGREE DAYS	1185	945	813	473	149	50	0	15	82	624	630	1441	6407
COOLING DEGREE DAYS	0	0	0	6	76	192	244	223	45	0	0	0	786
RH													
MEAN (PERCENT)	68	67	64	61	61	73	69	70	72	75	69	77	69
HOUR 00 LST	73	73	72	70	70	83	84	81	85	82	79	82	78
HOUR 06 LST	77	78	80	79	81	87	87	86	88	86	84	81	83
HOUR 12 LST	60	56	53	47	48	62	53	56	55	64	53	70	56
HOUR 18 LST	64	58	49	47	45	57	53	55	59	69	60	77	58
S PERCENT POSSIBLE SUNSHINE													
W/O NUMBER OF DAYS WITH:													
HEAVY FOG(VISBY <= 1/4 MI)	2	1	1	0	1	4	1	1	9	2	2	6	30
THUNDERSTORMS	0	0	2	2	4	9	6	4	2	0	0	0	29
CLOUDNESS													
SUNRISE-SUNSET: (OKTAS)													
CEILOMETER (<= 12,000 FT.)													
SATELLITE (> 12,000 FT.)													
MIDNIGHT-MIDNIGHT: (OKTAS)													
CEILOMETER (<= 12,000 FT.)													
SATELLITE (> 12,000 FT.)													
NUMBER OF DAYS WITH:													
CLEAR													
PARTLY CLOUDY													
CLOUDY													
PR													
MEAN STATION PRESS. (IN.)	28.10	28.11	28.01	28.01	28.02	27.95	28.05	28.05	28.12	28.00	28.09	28.06	28.05
MEAN SEA-LEVEL PRESS. (IN.)	30.14	30.12	30.01	29.97	29.96	29.87	29.97	29.98	30.07	29.97	30.08	30.11	30.02
WINDS													
RESULTANT SPEED (MPH)	5.3	2.1	1.5	3.6	2.2	2.8	1.2	1.9	2.4	3.1	2.2	7.1	1.1
RES. DIR. (TENS OF DEGS.)	30	31	21	05	24	05	13	17	14	32	28	32	32
MEAN SPEED (MPH)	11.2	10.5	13.0	13.0	11.3	8.2	7.2	8.9	7.5	10.5	8.7	12.6	10.2
PREVAIL.DIR.(TENS OF DEGS.)	34	33	16	35	20	02	17	15	14	32	18	33	33
MAXIMUM 2-MINUTE WIND													
SPEED (MPH)	45	44	48	41	36	46	38	35	32	43	33	44	48
DIR. (TENS OF DEGS.)	34	24	22	35	34	33	36	31	34	30	32	31	22
DATE OF OCCURRENCE	12	09	23	05	15	06	16	04	27	02	25	25	MAR 23
MAXIMUM 3-SECOND WIND:													
SPEED (MPH)	53	55	60	51	45	66	49	44	43	55	41	54	66
DIR. (TENS OF DEGS.)	33	22	21	36	35	34	35	36	35	30	32	33	34
DATE OF OCCURRENCE	12	09	23	05	15	06	16	04	27	02	25	25	JUN 06
PRECIPITATION													
WATER EQUIVALENT:													
TOTAL (IN.)	0.30	0.88	0.14	2.56	2.05	8.27	2.70	2.40	0.96	3.39	0.17	1.76	25.58
GREATEST 24-HOUR (IN.)	0.13	0.56	0.07	1.12	0.92	2.35	1.66	0.64	0.42	1.19	0.13	0.59	2.35
DATE OF OCCURRENCE	25-26	13	22	29	12-13	05-06	24	26	11	21-22	23	08	JUN 05-06
NUMBER OF DAYS WITH:													
PRECIPITATION 0.01	8	3	4	11	13	21	9	9	5	15	4	7	109
PRECIPITATION 0.10	1	2	0	6	6	12	5	8	3	7	1	5	56
PRECIPITATION 1.00	0	0	0	1	0	4	1	0	0	0	0	0	6
SNOWFALL													
SNOW,ICE PELLETS,HAIL													
TOTAL (IN.)	5.4	7.7	0.2	2.3	0.0	0.0	0.0	0.0	0.0	4.5	T	26.5	46.6
GREATEST 24-HOUR (IN.)	2.0	5.5	0.2	1.8	0.0	0.0	0.0	0.0	0.0	3.8	T	9.7	9.7
DATE OF OCCURRENCE	23	13	31	05						10	16+	08	DEC 08
MAXIMUM SNOW DEPTH (IN.)	4	6	2	2	0	0	0	0	0	1	0	15	15
DATE OF OCCURRENCE	28+	14	03+	05						12+		27+	DEC 27+
NUMBER OF DAYS WITH:													
SNOWFALL >= 1.0	2	2	0	1	0	0	0	0	0	1	0	6	12

published by: NCDC Asheville, NC

2

Figure 2.
Grand Island LCD Meteorological Data for 2009.

Examine page 3 of the *LCD, Annual Summary* (**Figure 3**) entitled **"NORMALS, MEANS, AND EXTREMES"**. "Normals" are averages of individual weather elements over a fixed period of time, usually 30 years. Normals for this *LCD* were based on 1971-2000. "Mean" values are averages for the entire period of record of the weather element. Respond to the following:

16. The normal monthly temperatures (Normal Dry Bulb) range from a low of 22.4 °F in January to a high of [(*75.8*)(*78.5*)(*81.6*)(*87.4*)] °F in the month of July in Grand Island, NE.

17. The Highest Daily Maximum temperature ever recorded at Grand Island was [(*99*)(*107*) (*110*)(*111*)] °F in August 1983.

18. The mean number of days per year with thunderstorms is [(*23.5*)(*32.4*)(*45.2*)(*56.8*)].

19. Over the year the mean wind speed is [(*11.2*)(*11.8*)(*12.5*)(*13.8*)] mph from a direction coded as *17* (tens of degrees measured clockwise from north), meaning essentially from the south. [Wind from due south would be coded as *18*.]

20. The normal yearly total precipitation is [(*17.44*)(*21.52*)(*23.22*)(*25.89*)] inches.

21. This is [(*much less than*)(*nearly equal to*)(*much more than*)] the total for 2009.

22. Also included as part of the *LCD, Annual Summary* is a brief narrative describing the location and climatic aspects of the area surrounding the local NWS office. According to the Grand Island description in **Figure 4**, its climate is described as primarily [(*"maritime"*)(*"continental"*)] in nature.

23. Incursions of maritime tropical air from the Gulf of Mexico [(*do*)(*do not*)] make it to Grand Island.

24. Grand Island's east to west upslope terrain produces episodes of [(*dust storms*) (*blizzards*)(*fog and low stratus clouds*)] when winds are from the east.

As directed by your course instructor, complete this investigation by either:

1. *Going to the Current Weather Studies link on the course website, or*
2. *Continuing to the Applications section for this investigation that immediately follows in this Investigations Manual.*

NORMALS, MEANS, AND EXTREMES
GRAND ISLAND (KGRI)

LATITUDE: 40 ° 57'N LONGITUDE: -98 ° 18'W ELEVATION (FT): GRND: 1850 BARO: 1844 TIME ZONE: CENTRAL (UTC -6) WBAN: 14935

ELEMENT	POR	JAN	FEB	MAR	APR	MAY	JUN	JUL	AUG	SEP	OCT	NOV	DEC	YEAR
TEMPERATURE °F														
NORMAL DAILY MAXIMUM	30	32.6	38.6	49.5	61.9	71.9	83.0	87.1	84.8	76.9	64.6	46.8	35.3	61.1
MEAN DAILY MAXIMUM	109	34.2	37.1	49.4	61.9	73.1	82.3	89.5	87.3	77.2	66.3	49.2	37.3	62.1
HIGHEST DAILY MAXIMUM	63	76	79	90	96	101	107	109	110	104	96	84	76	110
YEAR OF OCCURRENCE		1990	2006	1986	1989	1989	1988	2006	1983	1998	1947	2006	1964	AUG 1983
MEAN OF EXTREME MAXS.	110	58.0	63.7	75.4	85.8	90.9	98.1	100.7	99.1	94.0	85.9	72.3	61.0	82.1
NORMAL DAILY MINIMUM	30	12.2	17.7	27.0	37.8	49.3	59.1	64.4	62.3	51.8	39.3	25.9	15.9	38.6
MEAN DAILY MINIMUM	109	12.9	16.3	26.4	37.5	49.3	58.4	64.7	62.4	51.8	40.2	26.5	16.9	38.6
LOWEST DAILY MINIMUM	63	-28	-21	-21	7	23	38	42	40	23	9	-11	-26	-28
YEAR OF OCCURRENCE		1963	1994	1960	1975	1967	1954	1971	1950	1984	1997	1976	1989	JAN 1963
MEAN OF EXTREME MINS.	110	-9.7	-4.1	5.5	21.7	34.3	45.7	53.0	50.6	35.5	23.1	8.5	-4.1	21.7
NORMAL DRY BULB	30	22.4	28.2	38.3	49.9	60.6	71.1	75.8	73.6	64.4	52.0	36.4	25.6	49.9
MEAN DRY BULB	109	23.6	26.7	37.9	49.7	61.2	70.4	77.1	74.9	64.5	53.2	37.9	27.2	50.4
MEAN WET BULB	26	21.6	24.7	33.2	42.7	53.7	62.4	67.2	65.8	56.4	44.5	32.2	23.5	44.0
MEAN DEW POINT	26	17.5	20.8	28.4	37.7	50.0	59.0	64.4	63.1	52.4	39.7	27.8	19.7	40.0
NORMAL NO. DAYS WITH:														
MAXIMUM >= 90	30	0.0	0.0	0.1	0.7	1.0	7.7	13.0	10.1	4.8	0.3	0.0	0.0	37.7
MAXIMUM <= 32	30	14.3	10.0	3.6	0.3	0.0	0.0	0.0	0.0	0.0	0.1	3.9	11.6	43.8
MINIMUM <= 32	30	30.6	26.7	21.9	8.1	0.5	0.0	0.0	0.0	0.5	6.3	23.1	30.1	147.1
MINIMUM <= 0	30	6.4	3.7	0.6	0.0	0.0	0.0	0.0	0.0	0.0	0.0	0.4	3.3	14.4
H/C														
NORMAL HEATING DEG. DAYS	30	1310	1031	819	452	175	23	3	7	114	401	843	1207	6385
NORMAL COOLING DEG. DAYS	30	0	0	1	11	48	218	349	285	107	8	0	0	1027
RH														
NORMAL (PERCENT)	30	71	71	68	64	67	65	68	71	66	64	70	73	68
HOUR 00 LST	30	75	76	74	72	76	75	77	80	76	72	76	77	76
HOUR 06 LST	30	77	79	80	80	83	83	84	86	83	80	80	79	81
HOUR 12 LST	30	63	61	57	51	55	52	55	57	51	49	58	64	56
HOUR 18 LST	30	67	63	55	49	53	49	53	55	51	53	64	70	57
S PERCENT POSSIBLE SUNSHINE														
W/O MEAN NO. DAYS WITH:														
HEAVY FOG(VISBY <= 1/4 MI)	46	1.8	2.3	2.3	1.2	1.0	0.8	0.9	1.8	1.4	1.9	2.0	2.5	19.9
THUNDERSTORMS	62	0.0	0.2	1.3	3.5	7.4	9.3	8.6	7.6	4.9	1.8	0.5	0.1	45.2
CLOUDNESS MEAN:														
SUNRISE-SUNSET (OKTAS)	1	6.4	6.4	5.6	6.0	4.8	3.6	4.4	3.7	3.6	4.3	4.0	4.8	4.8
MIDNIGHT-MIDNIGHT (OKTAS)	1	6.4	6.4	7.2	6.4	4.8	4.0	4.4	3.6	4.0	4.5	4.0	4.8	5.0
MEAN NO. DAYS WITH:														
CLEAR	2	5.0	7.3	7.0	5.5	9.0	12.0	11.0	13.5	7.5	9.0	9.0	10.5	106.3
PARTLY CLOUDY	2	3.0	4.3	3.7	5.5	3.0	4.3	7.5	4.0	2.5	3.0	4.5	2.5	47.8
CLOUDY	2	10.7	7.3	7.0	9.0	8.7	4.7	4.5	5.0	3.0	6.0	5.5	10.0	81.4
PR MEAN STATION PRESSURE(IN)	26	28.11	28.10	28.04	27.98	27.98	27.98	28.03	28.06	28.06	28.07	28.07	28.10	28.05
MEAN SEA-LEVEL PRES. (IN)	26	30.14	30.12	30.03	29.94	29.92	29.90	29.94	29.98	30.00	30.03	30.07	30.12	30.02
WINDS MEAN SPEED (MPH)	26	11.2	11.5	12.6	13.4	12.1	10.9	9.5	9.2	10.6	11.0	11.3	11.2	11.2
PREVAIL.DIR(TENS OF DEGS)	34	34	36	36	36	17	17	17	17	17	19	19	34	17
MAXIMUM 2-MINUTE:														
SPEED (MPH)	17	54	53	53	52	59	68	59	52	45	51	55	52	68
DIR. (TENS OF DEGS)		34	33	34	21	29	30	29	22	28	33	32	31	30
YEAR OF OCCURRENCE		1996	2002	1996	1994	2000	1993	1993	1999	2004	1996	2005	1997	JUN 1993
MAXIMUM 3-SECOND														
SPEED (MPH)	17	64	61	61	60	80	77	72	70	52	61	62	63	80
DIR. (TENS OF DEGS)		35	33	33	34	24	30	29	23	21	31	32	32	24
YEAR OF OCCURRENCE		1996	2002	1996	1999	1996	1993	1993	1999	2006	2008	2005	1997	MAY 1996
PRECIPITATION NORMAL (IN)	30	0.54	0.68	2.04	2.61	4.07	3.72	3.14	3.08	2.43	1.51	1.41	0.66	25.89
MAXIMUM MONTHLY (IN)	70	1.65	3.39	6.63	4.98	9.04	13.96	10.38	8.73	9.00	5.99	3.77	2.17	13.96
YEAR OF OCCURRENCE		1960	1971	1987	1999	2008	1967	1993	1977	1965	2008	1983	1968	JUN 1967
MINIMUM MONTHLY (IN)	70	T	0.02	0.01	0.09	0.43	0.50	0.22	0.50	0.01	0.00	T	0.02	0.00
YEAR OF OCCURRENCE		1986	1996	1967	1989	1964	1978	2003	1940	1998	1958	2007	1943	OCT 1958
MAXIMUM IN 24 HOURS (IN)	70	1.38	2.21	3.15	3.30	7.21	4.54	5.41	4.12	5.88	2.75	1.90	1.20	7.21
YEAR OF OCCURRENCE		1947	1971	1979	1964	2005	1967	1950	1977	1977	1968	1996	1968	MAY 2005
NORMAL NO. DAYS WITH:														
PRECIPITATION >= 0.01	30	5.7	5.6	7.6	9.0	11.3	9.3	9.4	8.0	7.0	5.9	5.8	5.2	89.8
PRECIPITATION >= 1.00	30	0.0	0.1	0.4	0.5	1.1	1.1	0.7	0.9	0.5	0.3	0.3	0.1	6.0
SNOWFALL NORMAL (IN)	30	6.2	5.9	6.8	1.5	0.*	0.0	0.0	0.0	0.0	0.2	1.0	4.7	32.9
MAXIMUM MONTHLY (IN)	69	17.5	21.5	21.6	9.0	4.5	T	T	T	3.8	9.8	17.1	26.5	26.5
YEAR OF OCCURRENCE		1993	1969	2006	1984	1947	2008	1991	1992	1985	1991	1983	2009	DEC 2009
MAXIMUM IN 24 HOURS (IN)	69	10.5	15.0	12.2	7.3	4.5	T	T	T	3.8	9.8	11.2	12.0	15.0
YEAR OF OCCURRENCE'		2002	1984	1984	2003	1947	1991	1991	1992	1985	1991	1983	1968	FEB 1984
MAXIMUM SNOW DEPTH (IN)	63	18	17	16	5	4	T	0	T	1	7	14	20	20
YEAR OF OCCURRENCE		1974	1969	2006	2003	1967	1952		1951	1985	1997	1983	1968	DEC 1968
NORMAL NO. DAYS WITH:														
SNOWFALL >= 1.0	30	2.0	2.0	1.7	0.4	0.0	0.0	0.0	0.0	0.1	0.1	1.5	1.7	9.5

published by: NCDC Asheville, NC   3   30 year Normals (1971-2000)

Figure 3.
Grand Island LCD Normals, Means, and Extremes.

2009
GRAND ISLAND
NEBRASKA (KGRI)

The city of Grand Island is located in the shallow Platte River Valley in south-central Nebraska, less than 50 miles from geographical center of the contiguous United States. The complex of the Loup River and its tributaries converge approximately 20-25 miles northwest to north of the city, then flows east across the state. The terrain immediately around Grand Island is flat, sandy, loam. Just to the north is the south boundary of the Nebraska Sandhills. The terrain slopes gently upward from the Missouri River valley in eastern Nebraska to the Rocky Mountains of Wyoming and Colorado.

The climate is primarily continental in nature with occasional incursions of maritime tropical air from the Gulf of Mexico and modified maritime polar air from the Pacific Ocean. Winter time outbreaks of cold, dry, Arctic air from Canada are also common, usually accompanied by strong biting winds.

The east to west upslope wind flow provides periods of fog and low stratus, while a west to east wind component provides a warm and dry Chinook wind effect. Dry season dust storms occur infrequently with these Chinook winds. These have been reduced in recent years by increased farm irrigation and soil management techniques. Growing season humidities have also been increased by the increased use of irrigation in farming. Summers are usually hot and dry with temperatures often reaching 100 degrees or more. Late spring and early summer is the peak season for severe thunderstorms with frequent hail and occasional tornados. Winters are punctuated by occasional severe blizzards with temperature variations that range from mild to bitterly cold.

Based on the 1971-2000 period, the average first occurrence of 32 degrees Fahrenheit in the fall is October 8th, and the average last occurrence in the spring is May 6th.

Figure 4. 2009 Grand Island, Nebraska (GRI).

Investigation 15B: Applications

LOCAL CLIMATE DATA

In addition to **Local Climatic Data, Annual Summary**, there is available from the National Climatic Data Center **Local Climatic Data, (Edited), LCD, Monthly Publication**. This publication contains daily information on a 3-hour basis for each month. A sample copy can be viewed at *http://www7.ncdc.noaa.gov/IPS/lcd/lcd.html*. (Adobe Acrobat Reader required.)

25. Most National Weather Service (NWS) offices post monthly local climatological data on their Internet website. **Figure 5** is a sample from Grand Island for February **[(*2007*) (*2008*)(*2009*)(*2010*)]**.

26. NWS offices reporting monthly climatological data typically follow the same format (WS Form F-6) as shown in **Figure 5**. For each day of the month, data concerning temperature, precipitation, wind, etc., are reported in a table. Following the table, additional information is usually provided. From the table, or from the summary shown below the table, it can be seen that during the month reported, the highest temperature (column 2, "MAX") was **[(*32*)(*38*)(*44*)(*52*)]** °F.

27. The average monthly temperature (given below the table, in second part of Figure 5 under [TEMPERATURE DATA]) was **[(*17.6*)(*23.8*)(*31.7*)(*46.2*)]** °F.

28. The total number of heating-degree days (bottom of column 6A, "HDD") for the month was **[(*65*)(*577*)(*946*)(*1146*)]**.

29. The total snowfall for the month (in column 8, "SNW", also in second part of Figure 5 under [PRECIPITATION DATA]) was **[(*7.4*)(*8.5*)(*14.6*)(*25.4*)]** in.

Visit the website of the NWS Forecast Office nearest you that reports monthly local climatological data in a format similar to that shown in this investigation's example. To find the NWS office near you, go to: *http://www.wrh.noaa.gov/wrh/forecastoffice_tab.php*.

Click on one of the locations appearing on the map. The NWS office website you call up will have a left-side menu which includes the term "Climate". Under "Climate", click on "Local" to view monthly summaries. Most NWS offices post the most recent monthly summaries covering the period of a year or more. Call up and print out a copy for the most recent full month reported for possible future reference.

Suggestions for further activities: The National Climatic Data Center provides many types of climatic data, a portion of which is provided from their website (*http://www.ncdc.noaa.gov/ oa/climate/climatedata.html*) free of charge. Under **Data & Products**, select "Free Data" to view those data provided via web access.

```
PRELIMINARY LOCAL CLIMATOLOGICAL DATA (WS FORM: F-6)

                                    STATION:   GRAND ISLAND NE
                                    MONTH:     FEBRUARY
                                    YEAR:      2010
                                    LATITUDE:  40 58 N
                                    LONGITUDE: 98 19 W


   TEMPERATURE IN F:          :PCPN:   SNOW:  WIND         :SUNSHINE: SKY      :PK WND
===================================================================================
1    2    3    4    5   6A  6B    7     8    9   10   11   12   13   14  15    16    17  18
                                            12Z  AVG  MX 2MIN
DY  MAX  MIN  AVG  DEP  HDD CDD  WTR   SNW  DPTH SPD  SPD DIR  MIN PSBL S-S  WX    SPD  DR
===================================================================================

1    32   20   26    2   39   0  0.01  0.2    4  7.6  16   20    M    M   10 18    25 320
2    38   12   25    0   40   0  0.00  0.0    4  4.4  12  340    M    M    2 18    14 120
3    35   19   27    2   38   0  0.00  0.0    4 10.4  17  190    M    M    6 18    21 190
4    35   31   33    8   32   0  0.26  1.3    4 11.0  18  140    M    M   10  1    25 140
5    31   27   29    4   36   0  0.09  1.2    6  9.0  16  360    M    M   10  1    18  20
6    33   21   27    1   38   0     T    T    6  3.2   7  170    M    M    8 128    8 160
7    34   29   32    6   33   0  0.05  0.8    6  3.5  13  360    M    M   10  1    16 360
8    30    9   20   -6   45   0     T    T    6 18.2  30  350    M    M    6         36 350
9    24    5   15  -12   50   0     T    T    6 16.1  25  330    M    M    2         33 310
10   25    0   13  -14   52   0  0.00  0.0    6  6.8  15  160    M    M    3  1    17 160
11   33   20   27    0   38   0  0.00  0.0    5  6.2  10  230    M    M    8  1    14 220
12   31   20   26   -1   39   0  0.00  0.0    5  6.4  15  350    M    M    8  1    18 350
13   40   20   30    2   35   0     T    T    5 10.5  28  340    M    M   10 18    33 330
14   23   14   19   -9   46   0     T    T    4 26.6  45  340    M    M    4         52 330
15   31   13   22   -6   43   0  0.00  0.0    4 21.8  33  330    M    M    6         44 340
16   33   16   25   -4   40   0     T    T    4 11.5  22  350    M    M    7         26 350
17   39   10   25   -4   40   0  0.00  0.0    4  5.4  10  140    M    M    1         13 140
18   36   18   27   -2   38   0  0.00  0.0    3  6.6  15   30    M    M    5         18  40
19   28   23   26   -4   39   0  0.07  1.0    2 10.7  17   20    M    M   10 18    21  30
20   26   23   25   -5   40   0  0.03  0.3    3  7.5  14   40    M    M   10  1    17  50
21   27   20   24   -6   41   0  0.14  2.5    5 14.1  21   20    M    M   10 18    24 360
22   27    9   18  -12   47   0     T    T    6  7.6  13  360    M    M    3 18    15 350
23   26    8   17  -14   48   0     T    T    6 12.7  26  350    M    M    5         31 350
24   22    0   11  -20   54   0  0.00  0.0    5  5.9  13  130    M    M    1         17 130
25   30    8   19  -12   46   0  0.00  0.0    5  9.1  16  160    M    M    1         21 160
26   44   20   32    0   33   0  0.00  0.0    5  6.4  13  350    M    M    1  1    16 330
27   31   22   27   -5   38   0  0.00  0.0    4  5.2  12  340    M    M    8 18    14  40
28   32   22   27   -5   38   0  0.01  0.1    4  6.9  13   80    M    M    9 128   16 100
===================================================================================
SM  876  459      1146    0  0.66      7.4 271.3          M        174
===================================================================================
AV 31.3 16.4                                9.7 FASTST   M    M    6     MAX(MPH)
                               MISC ---->  # 45 340                   # 52  330
===================================================================================
NOTES:
# LAST OF SEVERAL OCCURRENCES

COLUMN 17 PEAK WIND IN M.P.H.
```

Figure 5.
Preliminary Local Climatological Data (WS FORM: F-6) for Grand Island, Nebraska for February 2010
- page 1.

```
PRELIMINARY LOCAL CLIMATOLOGICAL DATA (WS FORM: F-6) , PAGE 2

                                        STATION:   GRAND ISLAND NE
                                        MONTH:     FEBRUARY
                                        YEAR:      2010
                                        LATITUDE:   40 58 N
                                        LONGITUDE:  98 19 W


[TEMPERATURE DATA]        [PRECIPITATION DATA]      SYMBOLS USED IN COLUMN 16

AVERAGE MONTHLY: 23.8    TOTAL FOR MONTH:   0.66    1 = FOG OR MIST
DPTR FM NORMAL:  -4.4    DPTR FM NORMAL:   -0.02    2 = FOG REDUCING VISIBILITY
HIGHEST:     44 ON 26    GRTST 24HR  0.32 ON  4- 5      TO 1/4 MILE OR LESS
LOWEST:       0 ON 24,10                            3 = THUNDER
                        SNOW, ICE PELLETS, HAIL    4 = ICE PELLETS
                        TOTAL MONTH:   7.4 INCHES  5 = HAIL
                        GRTST 24HR   2.5 ON 21-21  6 = FREEZING RAIN OR DRIZZLE
                        GRTST DEPTH:    6 ON 23,22 7 = DUSTSTORM OR SANDSTORM:
                                                       VSBY 1/2 MILE OR LESS
                                                   8 = SMOKE OR HAZE
[NO. OF DAYS WITH]        [WEATHER - DAYS WITH]     9 = BLOWING SNOW
                                                   X = TORNADO

MAX 32 OR BELOW:  17    0.01 INCH OR MORE:   8
MAX 90 OR ABOVE:   0    0.10 INCH OR MORE:   2
MIN 32 OR BELOW:  28    0.50 INCH OR MORE:   0
MIN  0 OR BELOW:   2    1.00 INCH OR MORE:   0

[HDD (BASE 65) ]
TOTAL THIS MO.  1146    CLEAR  (SCALE 0-3)   7
DPTR FM NORMAL   115    PTCLDY (SCALE 4-7)  12
TOTAL FM JUL 1  5325    CLOUDY (SCALE 8-10)  9
DPTR FM NORMAL   405


[CDD (BASE 65) ]
TOTAL THIS MO.     0
DPTR FM NORMAL     0
TOTAL FM JAN 1     0    [PRESSURE DATA]
DPTR FM NORMAL     0    HIGHEST SLP 30.38 ON 16
                       LOWEST  SLP 29.89 ON 13
```

Figure 5. Preliminary Local Climatological Data (WS FORM: F-6) for Grand Island, Nebraska for February 2010 - page 2.